What Consciousness Is

Third Edition

Karl Sipfle

To Elizabeth Sipfle. Esq., partner of our “firm”

Contents

Part I: Consciousness, The Full Stack: STFC + STHC

How Brain Makes Mind: The Principles of Operation (Part I)

Karl Sipfle*

Abstract

We present the principles of operation by which a brain makes a mind, at all scales necessary to cover the whole problem. We inventory the necessary capabilities for a mind. We divide conscious mind into four layers of increasing elaboration. For the principles of operations of the lower two layers, we provide the STFC theory. For the upper two, we provide the STHC theory. We survey the evolutionary progression from first twinge of experience to human capacities. We explain the types of memory and problem-solving we carry and by what structures they are made to happen. We compare to prior works and review the philosophical implications and stance. All of this is done with minimal incoming assumptions, and those made are declared.

Part I of this four-part article includes: 1. Introduction; 2. All the Elements of Mind; 3. A Multiscale Problem; 4. Definitions; 5. Important Assumptions; 6. Levels of Consciousness; 7. The Four-Layer Architecture: The Gross

---

*Correspondence author: Karl Sipfle, Independent Researcher (also working independently at NASA GSFC).
E-mail: ksipfle@umich.edu

Architecture of Consciousness; 8. Consciousness Architecture Layer 1: Fundamental Consciousness; 9. Sites of Fundamental Consciousness in the Brain; 10. The Nature of Sentience; and 11. Consciousness Architecture Layer 2: Sentience.

Keywords: Consciousness, physics, evolution, feeling, qualia, mind, cognitive, affective, sensation, memory, learning, attention, perception, recognition, decision-making, problem-solving, coordination, self, symbol manipulation, language.

1. Introduction

This paper comprises a Concept of Operations and Architecture Description Document for a mindful brain, plus rationale and process notes on their derivations.

The term "consciousness" has been used to address everything from the smallest fundamental capacities enabling sentience to the internal world of a human mind. In this paper, we address all of it.

We refer to an Architecture with which a mind is built. It is a layered architecture, cognizant of the building process of evolution.

The architecture we populate with theory. The theories are expected to see more revision over time than the architecture.

By the time we get up to vertebrate brains (appearing in Layer 3) we start to have a great deal of extant data on how lower mechanisms are exploited and can use it here.

The paper's scope is of natural brains rather than artificial ones, though it is relevant to the latter. The most elementary facts of brain anatomy and operations we don't belabor and are widely available (Kandel, 2021).

2. All the Elements of Mind

This is the complete list of ingredients for mind:

Reflex (underpinning mind)

Regulation

Sensation

Memory, of half a dozen different types, enabling Learning

Feeling

Prioritization

Attention, bottom-up and top-down

Perception, Recognition

Prediction

Action Planning

Imagination

Decision-Making

Problem-Solving

Coordination

Unity

Making sense

Differentiating the real from the recalled or the imagined

Self

Symbol manipulation

Language

A theory that describes how all of these things work is a complete theory of mind.

3. A Multiscale Problem

The biggest issue of the consciousness mystery is not self-reference and certainly not cognition, but the existence of feeling. This topic has been rightly called the Hard Problem (Chalmers, Facing up to the problem of consciousness, 1995).

There is the matter of why and how we can feel at all, what is "subjective experience." And there is why given the useful valuation that the feeling phenomenon provides, we should also actually feel.

The answer to the Whys appears to be that any value mechanism would do, and real pain and pleasure are usable natural phenomena that are here in this universe for evolution to employ.

There is also the What and the How. What is feeling, scientifically? We address that.

The next puzzle is as to how minds are woven from basic mechanisms. While our knowledge is not complete, much of this puzzle is easier.

Multiscale models (what Rolls (Rolls, 2016) call "levels of explanation") of less complicated systems have been employed for decades. To understand the whole of consciousness (or even just some important parts) it is necessary to examine it at multiple scales.

4. Definitions

Feeling: Experience. That which makes the Hard Problem (Chalmers, Facing up to the problem of consciousness, 1995).

Raw feeling: The concept of basal, indivisible feeling.

Qualia: The allegedly rawest feelings discernible to a mind, discovered through introspection.

Consciousness: Feeling and, typically, information processing affecting each other in organized fashion to make action decisions in the interests of a species.

Levels of Consciousness: A progression of Fundamental Consciousness, Sentience, Animal Consciousness, Human Consciousness (Sipfle K. , 2018), see below.

Fundamental Consciousness (FC): The smallest element of what distinguishes consciousness; the actual smallest bit of feeling. Pure pain or pure pleasure. Somewhat similar concepts have been called pre-consciousness, protoconsciousness (Penrose & Hameroff, Consciousness in the Universe, 2017), microfeels (Poznanski & Brandas, Panexperiential materialism: A physical exploration of qualitativeness in the brain, 2020), and fundamental feelings.

Mind: An island of Animal or Human Consciousness. On Earth, each requires and belongs to at most one brain.

Cognitive: related to thought and/or information processing and not to feeling (though in practice feelings typically *accompany* the cognitive).

Cognition: Thought, as opposed to feelings.

Perception: Informational understanding (not emotional).

Affective: related to feeling, not cognitive.

Valence: Goodness/positivity or badness/negativity.

Emergence: A consequent complex whole arising from the interactions within a system, that takes on its own characteristics not describable/expected in terms of just the individual elements of the system.

5. Important Assumptions

This paper makes a few assumptions, which are out of its scope to justify (instead see (Sipfle K. , The Nature of Fundamental Consciousness (preprint), 2018)).

Postulate 1: Physicalism is necessary and information- and computation-based theories cannot provide for the source of conscious experience.

Postulate 2: Consciousness functions within the same basic rules (including mathematics) as all other phenomena that also exist in Nature.

6. Levels of Consciousness

We start by conceptually sectioning the vast scope of what is called "consciousness" into a few large categories of increasing capability (adapted from (Sipfle K. , 2018)).

Level 1 Consciousness a.k.a. Fundamental Consciousness a.k.a. Base Consciousness

> This consists of the smallest elements that represent experience. (This is where free-standing experience occurs, without an observer.)

Level 2 Consciousness a.k.a. Sentience

> Consists of everything that must be added to level 1 to make the sentience of the simplest sentient organism. (This is where "subjective experience" occurs- it introduces the concept of an observer.)

Level 3 Consciousness a.k.a. Mind

> Contains everything above level 2 that enables the behaviors and experiences up to the most complex of nonverbal organisms (by adding cognitive componentry).

Level 4 Consciousness a.k.a. Human Consciousness

> That which has been attained by humans (by adding the symbolic manipulation that is used for language) (Edelman, 1993) (Rolls, 2016).

Level 1 Consciousness plus Level 2 Consciousness are the foundation beneath the entirety of "consciousness" defined as "a bubble of experience" if that bubble "consists of colours, sounds, smells, tastes, etc." (Gamez, 2018). Level 1

Consciousness has its own elemental bubbles of experience, which we may call "psybits (as Eccles did)," but the experience is more primitive and not in all occurrences integrated into any kind of mind or substantial piece thereof.

We will see that Level 1 is the root of actual experience, rather like Layer 1 of a computer network, the Physical Layer, moving around real individual electrons. In both cases, all the rest sits on top and is connected assemblies and software built on Level 1. If we simulate Consciousness Level 1 and Level 2 and then put all the same algorithms atop them as we find in natural brains, we can get all the same behaviors, but that "mind" will not *actually* feel.

Level 1 Consciousness is the fundamental consciousness that is the physics manifestation of what is needed to build, from it, consciousness as we know it. The nature of this low level and the connection from it upward answers Chalmers's Hard Problem: feeling is a phenomenon of our universe that preceded minds and was discovered and exploited by evolution to build minds. Once we put the horse before the cart, the Hard Problem evaporates.

IV **Verbal Mind.** Abstracting Wiring and "Algorithms"
Cognitive

III **Mind.** Wiring and "Algorithms"
Cognitive

II **Sentience.** Qualia STFC
Affective + Cog

I **Fundamental Consciousness.** STFC: Particles, Fields
Affective

Figure 1: Levels of Consciousness

7. The Four-Layer Architecture: The Gross Architecture of Consciousness

Next, we move from conceptual and abstract observations to the question of how brains and minds are actually built. We construct as Nature did, in layers whereby each earlier layer provides a platform for the next and new things become possible. Our identification and classification of Levels of consciousness corresponds well, it turns out, with layers of architecture of real minds. These are shown in Figure 4: Architectural Layers of Consciousness.

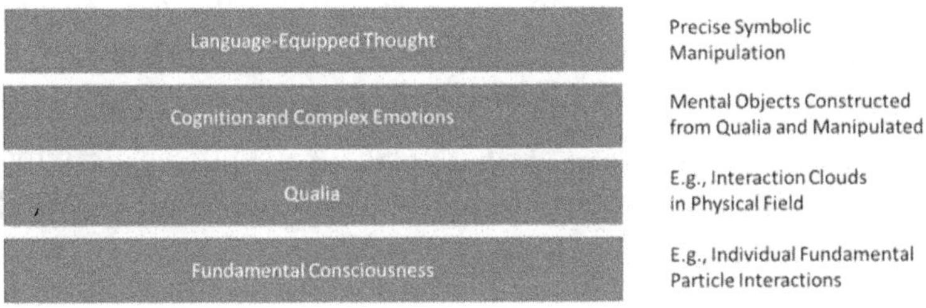

Figure 2: The Four Architectural Layers of Consciousness

These architectural layers of mind include, from the top down:

Language, integrated into thought.

Cognition, integrated with feeling.

Higher manipulations

Manipulation of base objects

Qualia, simple and complex.

Fundamental physics, whence come base feeling and objects that can be used to record information.

This we call the Four-Layer Architecture or FLA (colloquially the "Florida Architecture").

Now that we have specified the architectural concept of this evolved layering of consciousness, we shall discuss each of Layer 1, Layer 2, Layer 3, and Layer 4 Consciousness. In the process we also present our full-sweep ToC (Theory of Consciousness), comprising two sub-theories. The compliant and motivating theory described for Layer 1 and Layer 2 is the Sentonic Theory of Fundamental Consciousness, STFC. Strictly speaking, STFC is a metatheory, as it is agnostic on some specific aspects for which there are multiple candidates. The compliant theory described for Layer 3 and Layer 4 is the Systems Theory of Higher Consciousness, STHC. STHC is a synthesis drawing on the latest and best insights from neuroscience and theoretical and experimental psychology.

8. Consciousness Architecture Layer 1: Fundamental Consciousness

All phenomena we humans have identified have been seen to emerge from things that appear at the fundamental particle level through, maximally, the molecular scale.

Though we are not yet certain of the exact mechanism in Layer 1, we can characterize necessary key properties of it (and thus also speculate on actual mechanisms). This allows us to describe, without impediment because it is supposed to be more abstract, the nature of Layer 2.

Problems that will be faced in getting up to qualia (Layer 2) include

1. Qualia are not all of the same character.
2. Qualia are multitudinous.
3. A feeling is fused into a whole, and more intense feeling is a stronger whole. This is the "combination problem" first identified by William James, with many prior attempts to answer it (Harris, 2020) (Chalmers, The Combination Problem for Panpsychism, 2016).
4. At the bottom there are no Observers (in the mind sense), only observers (in the physics interactions sense) (Sipfle K. , The Nature of Fundamental Consciousness (preprint), 2018).
5. There must be a way- a path- for our minds to have evolved

We now discuss Layer 1 from the standpoint of what it must be like in order to support an explainable Layer 2.

Layer 1's province is the true indivisible, smallest feeling that can occur in the universe. Pursuing that leads us to some important (and non-traditional) conclusions.

First of all is the problem: for a feeling, who is feeling it? The process of drilling down to physics reveals that the question itself is essentially a meaningless assumption when we get to physical fundamentals rather than psychological concepts. When two electrons collide, which one is the observer?

> Feeling in its simplest form does not happen to an observer, it simply happens, as a freestanding event. The whole notion of a subject to experience something is a much higher development (and a true emergent phenomenon).

The Observer as we know it comes late in the upward progression. The Observer amounts to one clot of feeling and computation taking as input another clot of computation (and often, feeling). The Observer is a swelling in the unified corpus of mind and is not even 100% separated from the (internally) observed or 100% the same thing from one moment to the next.

At the bottom there is no observer, at least not one that is somehow fundamentally separate from the observed. Electrostatic interactions are not Observations, in a mental sense, they just happen. Fundamental feeling is the same way.

Secondly, our physics describes very well nearly everything in the so-called "physical" world, from the tiny to the immense and everything in between, while not at all describing such things as pain or pleasure. This tells us that

Postulate 3: Something *basic* is missing from, and must be added to, our physics, which is our description of reality.

Our feelings are real. Any Theory of Everything that really is that must include feeling or a way to make it. That the problem may be hard to solve does not make it invalid or meaningless as a problem. (Ironically, the only reason people study and develop physics in the first place is that they *want* to.)

The missing simplest physical feeling event we call Fundamental Consciousness. These are “psybits”; each functions as a “microquale” ((Poznanski, Theorising How the Brain Encodes Consciousness Based on Negentropic Entanglement, 2019) refers to somewhat similar “preconscious microfeels”).

Notice that there is not necessarily anything especially quantum about this. But because the fundamental level of Nature, which is the quantum level, extends up to the molecular scale, the exact process of feeling may conceivably occur up at this scale (at the maximum) rather than the particle scale. The Architecture, which is a framework or meta-solution, makes no demand that precludes that.

8.1.The Quale Garden

The first thing we should note as to the emergence of qualia from physical fundamental consciousness is whether there is any difference between the two. The answer is Yes, and this is terribly important to understanding of mind. In fact, there is also a large difference in nature between the different qualia.

If a quale for Red really were fundamental in the universe, then there would have to be hundreds or more of qualia for all the other cited experiences- the taste of wine, the sound from a musical instrument and so on. Quite obviously this is literally unnatural. The grand “plan” in every case to date has been found to be a few simple fundamental elements, then placed in combination. This means the typical quale is actually a composite structure, that is, the "atoms" of feelings of the mind are not the "atoms" of feeling of the universe. So, we do need something smaller than a quale, or alternatively, some of the qualia are more elementary than others. At bottom will be the smallest experiences in the universe (the psybits).

The usually cited qualia include a cast that are deeply dissimilar, not peers.

Of special note are the differences between, for example, Red and Pain. First of all, there are a great many qualia (especially in principle) just as "elementary" as Red, which immediately suggests it is not a truly elementary feel. Second, there is nothing special about Red from the standpoint of the universe; again, it is just one arbitrary spot in a large field of characteristics (even though as it happens our nervous systems are specifically sensitive to red light). Third and most saliently, Red has no intrinsic valence; Red is not intrinsically a painful thing or a bad thing or a good thing, it is just a factual condition.

Pain is entirely different in all three of these ways. There is pain and there is pleasure and there is nothing else in that family. Pain (or pleasure) is a very special occurrence in the universe, different from others. It has valence, and furthermore no factual content, only valence. The view that pain and Red are atomic siblings is false. Red and Blue and Shrill and Motivated are all of them more complicated than Pain and have internal structure. *To a mind*, or particular minds, they may be the simplest and most individual things detectable at the bottom, but that does not mean they are not made from simpler things (and they are).

In fact, the only clearly elemental qualia are pain and pleasure, the "emotional" ones (Sipfle K. , The Nature of Fundamental Consciousness (preprint), 2018). It turns out this is a profound observation and an important clue. Pain and pleasure are must-haves in the base feeling repertoire; no others are, they are constructed. The key categories of qualia, then, are Valent or Not, and Level of Complexity.

If a quale is a feel, then the simplest quale is the simplest feel. The simplest feel contains nothing other than feeling. This means a simplest quale cannot be about something, for then you have the feeling plus the thing about, and so here we encounter a difference in structural levels (levels of complexity).

To a mind, Red may seem as elemental as pain, because inside a mind you are sitting on top of everything needed to begin making one. This does not mean Red is as simple as pain inside the fabric of that platform. Indeed, this has been one of the primary confusions historically in understanding consciousness.

Even at the lowest level, one can't get from an "Is" (information) to any kind of "Ought" (which regards value and therefore feeling); the two different elements must become associated. The connection of feeling to information flows is ultimately what makes ideas "register." The idea is cognitive, the registration is feeling.

8.2. Necessary Qualities of Fundamental Consciousness

Qualia are the molecules of the mind, but not its quarks. And since qualia must be constructed of something similar (for workable emergence (Sipfle K. , The Primary Pitfalls on the Road to Understanding Consciousness, 2021)), this constructability becomes an essential feature of a successful architecture.

The answer is provided by the physical *field*, the mechanism that addresses another key problem, the fusion of feeling. This may be either an actual physical field, which we presume is the simplest explanation, or in principle another mechanism that is different but very similar in character.

We address first the build-up from fundamental pain to quale-level (mind-level) pain.

Note that by pain we mean "the painfulness of pain." Pure pain is the dysphoric ("emotional") aspect, the negative experience that is the end result of everything that causes pain.

If you are in slight pain and it gets much worse, we know that you will be experiencing the effect of more neurons activating, not just one or two becoming tremendously more active. At a physics level, the effect is many more psybits (fundamental feeling events) occurring, generally in close quarters. But what you *feel* at the mind level is not 1,000 individual small dots of pain. Your mind experiences the *same, unified thing, greatly amplified*. This means there must be a mechanism adding and joining these psybits into the whole pain. Note that simply interconnecting these feeling spots with neural information signals is not enough- that would still be separate little feelings, possibly now synchronized in time but still isolated as feelings and linked only by informational bridges. An additional mechanism is needed.

Physical interactions are all the behavior of forces. In a force field, every spot is subjected to the influence of every other spot in the same field. What results is a three-dimensional (four, including time) intensity cloud of a specific compositional shape. With this fusion, the otherwise freestanding feels join into a larger cloud of feeling. This is why you feel one big pain instead of a thousand little ones.

Pain is special as a feel/quale because pain is pain, on a small scale or a larger scale. Pain flows up from Layer 1 to Layer 2 in a rather direct and simple fashion.

8.3.Hypotheses Contained in STFC

STFC Hypothesis 1: Fundamental feeling is freestanding and requires no separate feeler. At most there is a physical interaction between fundamental elements.

STFC Hypothesis 2: There is a physical process that is fundamental feeling.

STFC Hypothesis 3: The fundamental feelings are pain and pleasure.

STCF Hypothesis 4: There exist pain and pleasure in minimal discrete bits independent of any brain, at the particle scale.

These are discussed in (Sipfle K. , The Nature of Fundamental Consciousness (preprint), 2018).

Re layer 2 we will discuss the informational qualia, which are very different. For example our Pain *that is distinctly located* also activates primary and secondary somatosensory cortex. But the same mechanism we have been discussing still underlies these pains. A field mechanism is needed in the consciousness architecture.

9. Sites of Fundamental Consciousness in the Brain

Let us now drill down on the matter of the most likely sites of action for consciousness in the brain.

This question is actually three: which neurons hold the sites of action, which regions of the brain contain these neurons, and where in the neurons is the site of action. For the first two questions- the macroscopic world of the cortical region and the cell- the evidence is stronger. For the third- the sub-neural, molecular scale (where physics lives), we are further away from knowing.

Macroscopically and microscopically, some areas of the brain show no signs of consciousness. This suggests that certain neurons and not others have evolved conscious (feeling) capability, directly causing pain or pleasure. This means the former have evolved nontrivial, nonchaotic access to the sentonic field.

9.1. Macroscopic Scale

It is medically known that human consciousness requires only

1. Cerebral cortex ("for awareness")

2. The Reticular Activating System of the brainstem ("for arousal")

The RAS projects to all cortex but mainly to prefrontal cortex.

If we look at limbic brain tissue as pain/pleasure-related, we find that it omits the stellate cells of cognitive tissue. This suggests that the feeling neurons ae pyramidal. Throughout the cortex pyramidal cells are the "workhorse" cells, the primary (and evolutionarily long-standing) units of the cortex, while the others provide regulation, lateral inhibition, relay, and pattern preprocessing to the deciding and publishing cells, the pyramidals. Closely associated with pyramidal cells, however, are chandelier neurons, whose axons always synapse exclusively on the axon initial segment of pyramidal cells. Therefore, chandelier cells may play a role in consciousness.

Pyramidals appear especially in cortical layers III and V so one (or both) of these is probably the most conscious layer(s). The III pyramidals and V pyramidals differ, with the latter being larger, and as we shall see, are used for different processing.

Layers II and IV of neocortex contain many stellate interneurons which are evolutionary newer, added once consciousness was in place.

Pain (dysphoria) centers can be found in the anterior insular cortex and anterior cingulate cortex (discussed later). The ACC is unique in its abundance of spindle cell neurons, which are connected to the anterior insular cortex and which are only found in very intelligent mammals. The spindle cells may not be themselves a site of consciousness, and do exist in other places, but their high density here may highlight the special importance of this area for very smart, very aware animals and the importance of pain to successful mentation.

In pleasure (euphoria), the insular cortex and orbitofrontal cortex are probably involved.

9.2.Molecular Scale

Anesthetic experiments and pathology and injury studies (e.g., (Hemmings, 2019)) indicate that it is not receptors directly and it is firing sequences that cause consciousness (and also that some neurons make consciousness and others do not). Most likely, spike generation or transmission is related to consciousness by a very short causal path.

Consciousness appears to occur in axons. The charge travel pattern of a neural spike is complex, unusual, and dimensionally precise. Here we have a toroid of deluges of ions across the axonal membrane, itself traveling along the considerable length of the fixed-diameter axon.

This means the ion rush through these channels (Fain, 2014) may be the site of consciousness.

While dendrites often taper, axons maintain a constant radius. The exact diameter varies by neuron type. It is typically about one micron. No neuron ever has more than one axon (though it may branch extensively). The axon initial segment consists of a specialized complex of protein molecules.

Note that the physics of axonal spikes might be replicated by a constructible apparatus for experimental purposes. This would allow for more intense examination than is readily possible in a brain or even brain slice. (Effects discovered could then be sought in real brain tissue.)

Microtubules have been suspected in consciousness theories (Penrose, The Emperor's New Mind, 1989). Much like axons but at a smaller scale, they are of even more constant and small radius and run the length of the axon and whole neuron. They experience the toroidal voltage pulse running down the axon that possibly induces something in or on the microtubule. These precisely dimensioned microtubules might function as resonators of some kind. In our model they would produce resonances in the sentonic field, or of a causative agent. Subsequent spikes would regenerate a decaying resonance. Also worth noting is that many differing molecules bind to microtubules.

It is not clear yet what the exact mechanism is, but it is clear that there are several possible pathways for what is needed, which is ion control of sentonic events and vice versa. At a slightly higher level knowing that is all we need, and we can await the elucidation of the details.

The sentonic field provides another means, requiring no direct cabling, for points in the brain to communicate with each other ("ephaptically") (Sipfle K. , Support for the Sentonic Theory of Fundamental Consciousness (STFC), 2019).

While cycling (and resonance) of signaling in the recurrent connections in neural networks may provide a regenerative effect that fortifies consciousness (and provide computation and correlation), it is not the base mechanism (cycling is a base mechanism of short-term memory, however).

Layer 2 takes us from individual glints of minimum feeling to our first unitary body of larger feeling. It provides the combining and fusing. It is the weaving of the elementary and separate into whole cloth.

This layer provides fusion, minimum awareness, and discernability between individual fused configurations.

This layer exists in physics, possibly chemistry, and brain biology to provide the needed surrounding conditions and location in the real world.

In keeping with Layer 1, there are two conceivable mechanisms providing this.

The first is the field. The nature of fields is such that each point is influenced by all neighbor points. Electrons for example form a summary, mutually owned electric charge profile owing to their relative locations. Sentonic charge and its local interactions would presumably have similar field effects.

The second possible contributor at this scale is quantum entanglement at maximum scales up to perhaps a millimeter. A problem with this mechanism, though, is that it is unlikely to serve at brain-size scales, as the brain is a messy, hot environment that would not support any large-scale entanglement. It could be speculated that while large scale, stable entanglements are infeasible, brief chains might repeatedly appear. The earliest actualizations of this layer could conceivably rely on entanglement across macroscopic scales, for example on the order of a neuron.

This layer is the first in which we now have a feeling object larger than elementary ones, that can then be impinged upon by external elementary ones or others like itself. Now we have something we can call sentient- a unitary blob responding to impingement with macroscale feeling.

In other words, at this level something resembling a Feeler occurs, a Subject, whereas before there were only equal minimal partners in mutual physics interactions.

Feeling is fundamental to the large-scale operation of vertebrates and not necessary to the behavior of other life forms, so this layer probably organized a bit before vertebrates but not greatly before. So far, we have no evidence that a single-celled organism or an ant has any reason to feel.

In addition to fusion this layer likely contributes or sets up for persistence.

Important and subtle things happen in this layer. A composite feeling and sub pieces thereof must be not only fused but individually distinguishable (Sipfle K. , The Solvability of The Hard Problem of Consciousness and its Relation to Fundamental Consciousness and Human Consciousness, 2019).

This layer provides "molecules" of feeling from which real qualia can be built.

Note also here that you can have an *awareness* of something without it capturing *attention.* As humans we would call such things "at the periphery" (because our actual or minds' foveas have not been fixed on them).

10.1. Hypotheses Contained in STFC

STFC Hypothesis 5: Some biological brains have evolved to organize and exploit pain and pleasure.

STFC Hypothesis 6: The fusion of feeling across the brain occurs in a force field.

11. Consciousness Architecture Layer 2: Sentience

Unification of information can occur in one way by the interconnected firing of informational components, but that does not explain a feeling of Red that is both unitary and different from a feeling of Blue.

One will come to know a feeling of Red through very long experience with it, but as has been oft-cited, exactly what red is like to me may not be identical to what it is like for you. What is important in the fabric of mind is that it feels like *something,* and it feels *different* from Blue, and from early on we then take these things (feelings) for granted as our minds continue to build.

Separate occurrences are not enough. *Both for feeling and for informational meaning* (which suggests the same mechanism is at play), we must have both the individual sub-aspects and unification of those into a whole. Assuming that there is only one level- qualia- is one of the fundamental errors in traditional thought on the topic, even before we get to the assumption that these individual qualia are the same thing as physics acts rather than compositions of physics acts.

We are provided an excellent clue toward the solution of the problem with pain (and pleasure). Pain is all the same, at all scales. It contains no information; it is purely valent.

Once we understand the necessary nature of fundamental consciousness, and that this is not the same as qualia, the next question presenting itself is how do we get from the former to the latter?

11.1. Meta-Symbols

To make thinking requires unique "symbols," things we can manipulate that represent other things. Even a symbol is divisible, and that is even before we deal with a symbol having both a name and an appearance. What is needed to start are unique identifiers- tags (meta-symbols)- to associate with large clouds of meaning. The tags, which are themselves each feeling clouds, become associated with yet more clouds, for its name, for its written visual representation if any, plus possible explicit association with net feelings of goodness or badness.

This assemblage provides two things: feeling and unique pattern of real things in space. One such blob "feels different" from another, both in the sense of the blob doing its own feeling and of interacting blobs "experiencing" it (which is the underpinnings of mind, and which is two peers-being feeling- affected by each other).

The little clouds *feel different*. This, ultimately, is why Red is different from Blue and a hundred other "raw" feels. Red and blue must feel different, and we must have a means by which this can be so. This issue is right at the nexus between the cognitive and the emotional mental realms. It is not enough to be different informationally.

Different patterns of feeling in the clouds allow for the existence, recognizability, and reference of tags that are different. By themselves, they (the tags) have no meaning or clear symbolic identity, the purpose is simply to be different and distinguishable- unique identifiers. Precise visual or aural symbols can then exist- which themselves have some complexity- and then be attached to these tags, just as the many specific experiences that make up the meaning are attached to them.

What does "attached" mean? The attachment *to* both the symbolic detail and the meaning detail can be just cognitive-neural links that don't, themselves, have to feel. This is the connection between the what-it-is-like for the remembered meaning (and similarly for the visualization of the symbol) to the what-it-is-like of the tag.

These three identifiable clouds of experience (generic tag as a handle, symbolic name/description, and situational experience) will bump up against each other in the overall continuous train of feeling being created in the lower brain, while the higher has the parallel cognitive conversations, to whatever relative degrees the content has high cognitive content vs. high impact.

Each part of each cloud is being stimulated by a cognitive circuit (possibly by way of an additional connected neuron whose only function is to tweak the field), and then unified by the natural field(s).

Cognition evolves to where the differentiated and distinguishable tags become persistently associated with much larger "clouds" of cognitive activity. With these handles, entire ideas can be effectively referenced and manipulated.

This is what “meaning” is to us. “Meaning” as a term in common use means: 1) a collection of informational associations and 2) subjective experience of it. This is the blossoming of “what it is like.”

Information is patterns of things, as with an abacus, and in full reduction the only things needed are the feelings. Anything else is intermediary, inert in the sense of importance to the abstract theory and architecture, other than to help arrange the patterns (which is where the wiring of the brain and physical shapes of neurons themselves come into play).

Random variability in the structure of neural nets and neurons yields unique, distinguishable signatures of the clouds of psybits mutually influencing across space where these activations are happening in the brain. Red feels like something and Blue feels like something and they are different.

The full picture of this is feeling stimulated in the field by informational circuits and a return path (otherwise feelings would have no effect), and this occurs both very far down and at multiple levels, ascending with the size and sophistication of thoughts (bigger concepts will literally be larger than tiny concepts). There is a very intimate and deep back and forth of feeling what is known and happening and knowing what is being felt that forms the stuff of our minds (so much so that it has been difficult for us to tease these two strands apart).

It is necessary to have physics at work up to the point of constructing qualia. Qualia are the link from the actual to the abstract. Qualia are a conceptualization of the reality of feelings. With the "hardware" realization in place, we can manipulate and trade in the constructs conceptually, and the rest of the stack becomes "software" (and connective wiring). An important transition is made at this point: feeling is absolutely critical to how our minds work and this must be understood. Once it is, though, feeling can be considered a known aspect and discussed and used in an abstract way in recognizing a working architecture that relies upon it.

We note that feeling provides "information" of a sort, but with important difference from ordinary information. Feeling enables a key portion of the control section that guides the operation of the data paths and computation, and this is going on at all levels of the mind (though not in all portions).

The difference can be understood in terms of implications. Knowing of either kind can be *encoded* in information, which is then read so as to know what is in the message. The knowing elicited may be of a fact or of a direct pain/pleasure experience. The two are different, even though both may be represented and, in that way transported, by information (and even if information is the patterning of feelings).

Another way to see the distinction is: Given enough book knowledge to read and enough mental ability and time you might understand any fact presented. However, to know what an experience was actually like can only be successfully transmitted to someone who has had the same experience or similar experiences. Otherwise, one must resort to knowing factual features about the experience and cogitating about a described experience instead of being able to emulate personally the experience of the sender, to sympathize.

The top two layers of the architecture presented do not discuss physics, they discuss the structure of our minds. The bottom two layers explain what the fundamentals are for the top two layers to work with, and make the connection up to the bottom of what we can introspect to, and call for specific explanation to ultimately prove those things at that bottom are the atoms of mind, validating the assumptions made by the higher two layers.

Note that stratification into layers correlates nicely with the different disciplines; neuroanatomists' findings, for example, are especially useful throughout Layer 3, and harder to use in explanation beneath or above.

We have seen that cognition manipulates information and so is not of necessity related to the stuff that encodes it. Feeling is not about information and has everything to do with actual physical process. This means in its simplest form, mind needs nothing more for the stuff than the feeling, and the wiring that helps pattern the occurrences of the feeling; no other stuff is strictly required other than to maintain life and the mechanics and housekeeping of biological functioning.

While we are here in the discussion, we can point out that this leads (distantly) to a subbranch of the present thinking that is (much) more speculative but not devoid of merit, involving the belief by some that life itself involves the fundamental consciousness force. The primary rational support for this notion, which probably originates from a "transcendent" suspicion (as did also that we are related to animals and stars), derives from the realization that feeling had to be discoverable by evolution. One possibility, then, is that the feeling necessary for our world was discovered quite early by evolution and has been used to enable life processes that speak little (at least, obviously) to its experiential nature, just as electromagnetism and gravity have been used early and often. Note that base consciousness is expected by this paper to be variations in a field present everywhere. From the other direction, it is also possible that our particular region of space was particularly seeded not only with proto-life elements but with proto-consciousness elements.

11.2. Primitive Self

There is a self- the one- before awareness of self. When do things enter one's consciousness? When the chief, large, dominating (though shifting) cloud of feeling with embedded cognition that we call Me is affected by other clouds- when they come into effective contact because of out of a background of feeling noise there is a local strengthening that bridges into the primary Me. The Me is already a large cloud of assembled feelings, that has a (fuzzy) perimeter- a cloud that hangs together with only thin "mist" of disorganized consciousness between it and other little clouds.

To enter your conscious mind, you must feel what is entering. Not to feel is not to be conscious of.

To be conscious of is not the same as the core mechanism of attention, which is a cognitive (that is, unfeeling) mechanism. What you feel will in large part (the other part being cognitive things like difference) determine what the attention circuits bring into focus, and the cognitive things (and feelings, too, as objects of attention) then amplified rather than suppressed will in turn result in feelings, which will then affect attention, but the attention proper is a cognitive device. Feeling occurs in its usual independent and passive way yet determines and thus drives the direction of thought.

Split-brain studies show one self in each hemisphere of the brain (even when still connected, the informationally linked but separated hemispheres may have distinct emotional lives). This suggests a prediction about the field strength formula of the consciousness (sentonic) force: it should feature high influence in the micron-millimeter scale and poor influence at the centimeter-meter scale. That would also make sense from an operational utility standpoint: Goodness v. badness values are most useful when they summarize each of subsets of everything that is going on.

11.3. Qualia

Qualia are the essence of the Hard Problem. Qualia are at the bottom edge of human mind (which is a whole represented by intense interconnections both mentally and physically). The bottom of the mind is well above the physical level making it possible, that is, fundamental consciousness.

In analyzing this layer, pain and pleasure are interesting as subjective experiences that are included as basic experiences to every human mind, but that do not demand a substantial range of possibles, as qualia (such as colors), in general, do. Suffering and pleasure are distillable in pure form from the human mind, and there is only one essential member of each of those two experience classes (or just two of one class if you prefer), not a spectrum or family.

In contrast, Red is *not* a single raw feel. Red, and anything else we calmly notice, has *impact*, a feel shared by many other "raw" feels. Furthermore, it is a visual (possibly envisioned) experience, and additionally to those two facts, it is also red.

Impact is like pain and pleasure without the sign. Impact is neutral oomph. (Impact may simply be a composite of negative and positive.) Minimum awareness is naked impact.

While this quale- red- may be among the least at once that the mind can do, and in that way this supposedly elemental thing may be atomic to the mind, it is already clearly made of multiple things. Commonly mentioned qualia are not the rawest of raw feels; they are more complex.

Such a minimal, actual experience for us is likely to be an *assemblage* of some kind of that which applies at the elementary physical level.

How do we feel red and blue? A little piece of your brain will light, consciously, when blue light comes in. However, this piece in isolation would not influence the connected whole of consciousness that we call the mind; it would be an outlier. This piece of brain is connected closely with a piece of brain that activates for any such visual color input (these connections are either direct or effectively so). Now we have a mob operating in concert, the experiencer of blue, and the experiencer of visual color upon which other color experiencers converge. This explains both the specific and the more general experience which are happening together. Yet higher convergence is to vision as a whole (fusion). These elements of experience are integrated into a whole, yet with introspection we can detect that both a common visual awareness and a specific redness are occurring.

What then makes the activity of the red brain patch different from the activity of the blue brain patch? One can say they are separate and so can hold separate information but that is only half the answer. Furthermore, there is nothing at bottom to us that feels special about red or blue, yet they *feel different*.

The two patches of brain tissue will have different microdetails physically, which could cause differing blobs of fundamental consciousness actions (psybits) in space (with these psybits in contact with one another by virtue of their field). It is not important on this point what the exact arrangement within these blobs is, just that they differ. The degree of difference is yet more pronounced with fundamental consciousness coming in positive and negative flavors, because there is not only an amplitude shape in space but also a pattern of +/- psybits within the blob.

What is happening is that the detailed tapestry of individual spots (psybits) has different detailed composition, so the feeling of the little mob is different. This effectively gives differentiating identities to each particular feeling; the experience of each is a little different (these then become associated in groups at higher levels of the mind, into richer experiences).

In a signed theory, a typical glob will have *impact* but not pain or pleasure because the plus and minus psybits cancel each other out *overall,* but each *pattern made* of the individual psybits exists and is different.

(*Continued on Part II*)

2. How Brain Makes Mind: The Principles of Operation (Part II)

How Brain Makes Mind: The Principles of Operation (Part II)

Karl Sipfle*

Abstract

We present the principles of operation by which a brain makes a mind, at all scales necessary to cover the whole problem. We inventory the necessary capabilities for a mind. We divide conscious mind into four layers of increasing elaboration. For the principles of operations of the lower two layers, we provide the STFC theory. For the upper two, we provide the STHC theory. We survey the evolutionary progression from first twinge of experience to human capacities. We explain the types of memory and problem-solving we carry and by what structures they are made to happen. We compare to prior works and review the philosophical implications and stance. All of this is done with minimal incoming assumptions, and those made are declared.

Part II of this four-part article includes: 12. Brain Organization; 13. Brain Structures for Higher Consciousness; and 14. From Association to Thinking.

---

*Correspondence author: Karl Sipfle, Independent Researcher (also working independently at NASA GSFC).
E-mail: ksipfle@umich.edu

Keywords: Consciousness, physics, evolution, feeling, qualia, mind, cognitive, affective, sensation, memory, learning, attention, perception, recognition, decision-making, problem-solving, coordination, self, symbol manipulation, language.

12. Brain Organization

Before introducing architecture Layer 3 which will highlight the brain regions that enabled increasing capabilities in perception, cognition, evaluation, and motor planning, it is informative to take a brief, high-level tour of how the brain is organized.

Mammals have a hippocampal memory system (as well as others). This system cross references happenings across the entire brain. The hippocampus supports episodic memory. It is used also for location in body-relative terms in rodents, which are smell-dominant, and absolute location in space in the higher mammals, which are vision-dominant.

The brains of higher mammals have a ventral back to front cortical visual stream that is about Where and a dorsal back to front cortical visual stream that is about What.

All the outputs of informational and affective processing collide and result in decisions in the prefrontal areas (which are most of the human frontal lobes, which ultimately control movement), which along with the temporal areas, are highly developed in the highest mammals.

The outer gross layer of the human cerebrum, the cerebral cortex, consists of 16 billion neurons in 3 to 6 layers. Most of human cortex is the more complex 6-layer type.

The cortex features a hierarchical processing structure, with feedforward projections (axons) from one area of cortex to the next-higher processing area. The concepts represented in each sequential region are of increasingly higher scope. The forward projections run from the pyramidal neurons in layers II and III of a region to layer IV, the input layer, of the next higher region.

All of the senses feature processing through multiple regions and then a coming together to fused multimodal representations in yet higher regions. The output of these is used by the prefrontal regions.

There are also backprojections- feedback- from higher to lower region, from pyramidal neurons in the higher region's layer V to layer I of the lower region. These facilitate learning, recall, and top-down attention (Rolls, 2016).

Besides these data paths, reward values are also calculated in a hierarchical fashion (Rolls, 2016).

There is also bottom-up, non-deliberate influence on attention.

The thalamus internally features lateral inhibition, likely supporting attention. More on the thalamus shortly.

At multiple levels of processing is short term memory, all levels simultaneously active, in the form of signal-regenerating loops of neurons. This is where you keep what you're thinking about.

Bottom-up emotional impulse decision-making wrestles in neural competition with top-down intentional decision-making to produce a winning action decision.

When in deep sleep, humans have no consciousness and slow brain waves are detected (delta waves).

Having moved into in REM sleep, humans are essentially conscious, with fully formed emotions but nonsensical disjointed cognitive content (Sipfle K. , A Neurobiological Theory to Explain the Lack of Rationality in Dreams, 2021). This may resemble the daily conscious experience of some other mammals, who do not evaluate what makes sense. Some also respond more in the moment and quickly forget earlier doings.

Waking generally follows REM sleep. The border period may include sensible thoughts but a much slower experience of the passage of real time with therefore faster thinking. Shortly after waking is reported to be a less rigidly constrained and more creative period than found during the day, of done at the right point in the sleep cycle (Mountinho, 2021).

Compared to REM sleep, wakefulness EEG shows frontal and posterior areas to be more synchronized with each other.

These facts suggest that the discernment of "making sense" happens in the prefrontal and takes time (and input of course).

In general, the cortex appears to be relatively passive flow-through, under control of the thalamus, which determines gross cortical states as well as focusing attention, and the diffuse activating system in the brainstem, which determines gross activation level. It appears that consciousness lies broadly in the cortex, though not everywhere.

Firing neurons are the presumed evokers of consciousness. Among these the pyramidal cells are the most likely sites of consciousness (or perhaps cells in close association with them).

There appears to be an excitatory wave of neural stimulus, the gamma wave (with a period of about 25 ms), sweeping from back to front in the cortex, which is the direction of feedforward from lower to higher processing (van Kerkoerle, 2014). ANN experiments by the author demonstrated a utility in rhythmic stimulus waves in pushing neural selection networks over firing threshold for the strongest incoming summed stimuli. Alpha waves (with a period of about 100 ms) represent feedback in the opposite direction.

Prominent types of learning, all relying on the strengthening of synapse effectiveness, are by repetition to establish recognizers to comprise the repertoires, and value- (feeling) based learning against actual or expected results for the decision selection networks following the recognition networks.

The inhibitory cortical neurons have fewer synapses and few long-range connections and are thus viewed as being related to the local stability of the excitatory circuits. The excitatory circuits appear to perform the major calculations.

The circuitry of thalamus and cortex form recurrent neural loops of two types, serving attention (via the isothalamus of the thalamus) and general activation (via the TRN of the thalamus).

All senses other than smell (which directly enters allocortex, which is agranular) first enters the isothalamus area and from there travel to the cortical information input layer, layer IV, of the primary sensory areas of the cortex. The cerebellum is also treated in this way as an input organ. Forwarded informational results from the region's processing emerge from layers II and III and enter as input to (or close to) layer IV of the next higher processing region. This locality is a fundamental principle of the evolution and capabilities of neocortex. The sensory inputs are sent from the thalamus to the prefrontal motor area as well as the sensory areas. Individual pyramidal neurons learn combinations of their inputs for which to fire via modification of the transmission strengths of selected synapses at their inputs (Hebb, 1949).

The second thalamic area (TRN) makes more diffuse connections both with lower brain nuclei and with the cortex, and plays a regulatory and gross control role rather than carriage of information. The TRN part of the thalamus receives cortical signals from cortical layer VI. It takes much more input than it produces output.

The superficial cortical layers (II, III) are involved in the feedforward processing path of increasingly high-level constructs, in other words, the fundamental information processing of perception and cognition. The inputs to layer IV synapse on stellate cells there, whose outputs go predominantly to the pyramidals in the superficial layers II and III. These pyramidals have heavy recurrent (both directions) highly local (several mm) connections, and these synapses are plastic (trainable). This allows for separate memories using space, only a few mm apart. The locality also supports topological maps, which themselves support hierarchical processing.

The superficial pyramidal neurons provide the main driving input to the deep neurons in cortical layers V and VI. This is the primary relationship between the superficial and deep layers. This also shows that most input into layer IV goes upward into the superficial layers. Some outputs of those layers then drive the deep layers.

Output from layer V, in addition to going to the basal nuclei, go to the prior cortical region as feedback into layer I, and are believed to be involved in recall and top-down attention. Some superficial neurons (II, III) also send some backprojections into Layer I. In the primary motor cortex, motor neurons are layer V pyramidals and their output heads to ganglia in the spinal cord which then project to muscles.

Output from layer VI goes to the noninformational (regulatory) part of the thalamus and to layer IV, both to inhibitory effect. Thus, there is a thalamocortical tuning loop from thalamus to layer IV to layers II and III to layer VI to thalamus.

In sum, we see "the superficial layers being specialized for feed-forward processing involving categorization into discrete representations using competitive learning to discover new categories, based on the convergence of inputs from different parts of the preceding cortical area(s)" (Rolls, 2016) with more sustained firing. The deep layers, particularly layer V, are smoother spatially (more continuous in categorization) and faster, with more bursty firing.

As with other networks, the connectivity of the brain ("connectome") features areas of high connectivity and high interconnectivity among them. These hubs, of which there are three cortical and three subcortical on each side, are in these locations (Heuvel, 2011):

Superior frontal cortex. Plans actions.

Superior parietal cortex. Location of visual objects.

Precuneus. Medial parietal cortex. Recently expanded in evolution.

Hippocampus, of central importance to memory.

Thalamus.

Putamen, part of the basal nuclei. Coordinates movement.

12.1 Cortical Evolution

Evolutionarily old cortex, known as allocortex, first appeared in reptiles as a distinctly three-layered structure.

Six-layered cortex (isocortex) appeared in mammals with the older allocortex at its margins. The thalamus supports neocortex; there are no thalamic connections with allocortex.

Per Rolls (Rolls, 2016), the correspondence of layers between the old allocortex and new isocortex is

Olfactory	**Hippocampal**	**Neocortex**
allocortex	allocortex	isocortex
lay 1		I horizontal axons from layer III pyramidals and axons from other regions and dendrites of pyramidals and some layer II stellates
lay 2	CA3	II pyramidals and some stellates
lay 2	CA3	III pyramidals
	Dent gy	IV granular layer (stellate cells,

which are what take the input)

CA1 V pyramidals (sends backprojections, sends to pyramidal tract neurons)

VI pyramidals (sends feedback backprojections to noninfo thalamus)

Olfactory cortex (pyriform cortex) is completely agranular. Granular cells are the stellate neurons.

Layer I of cortex contains mostly just connections between neurons. A characteristic local coupled arrangement exists in cortex between self-exciting pyramidals exciting and being inhibited by self-inhibiting interneurons, denoted here by P -- i. All cortical layers (except I) internally contain this pattern.

The older allocortex (reptilian) has a basic structure as so:

(Layer I)

Layer II/III: P -- i

P exciting both neuron types in Layer V/VI

Layer V: P -- i

P exciting both neuron types in Layer II/III

The mammalian innovation, neocortex, resulted from these changes (Beuhl, 2015):

Adding new connections from II/III into V, both of which tend to quiet V:

Adding a new inhibitory connection between existing layers, from II/III i to V P, thus creating an interlayer P – I coupling from Layer V to II/III.

Adding a new excitatory connection between existing layers, from II/III P to V i.

Adding a Layer IV, and

Adding a full II/III P to IV i interlayer coupling.

Adding an inhibitory connection from IV i to II/III i.

Adding an excitatory connection from IV P to V P, which tends to excite V.

Adding a Layer VI (which talks to the thalamus), and

Adding an excitatory connection from VI P to IV i.

Breaking the connection directly from V P to II/III P and inserting VI P and IV P in that path:

V P to VI P to IV P to II/III P

Nothing was deleted.

The consequences of all this are that:

II/III masters V.

II/III's informational influence on V remains but V's informational influence back to II/III is now gated by VI and IV.

A new layer IV preprocesses incoming information and is tightly coupled to II/III.

A new Layer VI influences IV and projects to the thalamus.

The focus of the elaborations is on feedforward information processing.

13. Brain Structures for Higher Consciousness

The richly emotional sections of the brain, (literally) closer to the ground, are both separated from and connected to the (literally) higher and stoic, calculating portions of the brain.

13.1 . Theory

In our theory, actual consciousness does not arise directly from resonant signal loops, but from the consistent stimulation of the sentonic field that such loops provide, by repeated operation of the basic stimulation action probably associated with neural firing. The specific resonant frequency is unimportant for this, it is whatever the brain loops happen to settle to.

The macro-sized consciousness is also thus probably not a quantum effect, even though the individual fundamental feelings operate at a quantum scale.

It has been theorized that neural synchronization accounts for the neural basis of consciousness (Crick, 1995). At a fundamental consciousness level, no, it doesn't. At a concentration of consciousness level, yes. At a much higher level involving attention, yes, by locking on to what is of interest. A competitively selected, regenerating loop provides a lock on the item of interest.

First, basic cognitive ability was evolutionarily achieved through the sequentially performed neural triad of candidate recognition, candidate ranking, singular final selection. For primary input and action, the first and last are easily done by relatively primitive means. The second is where the circuitry elaboration probably occurs in high-end brains, as this is where the scoring mechanism is applied in the competition among proto-ideas. This was critically enhanced by adding new cortical areas, each repeating the triad in a sequence of ever-higher-processing areas.

We know that the thalamus appears to gate, and thus determine to what is paid attention. It receives input from cortex, which is likely elaborate conclusions, in making those gating "decisions." The gating thalamus and the analyzing cortex cooperate, settling on what to settle on, which is what is found by cortex to be most interesting, which will be what produces the most related activity of inputs and memories and desires. This decision-making as to attention is probably another cortical takeover of functionality that used to be done by the thalamus alone. This takeover then enabled intentional and deliberate acts of attention.

13.2. Conjectures

Conjecture 1: Cortical columns/modules may exhibit a highly independent consciousness, co-influenceable like the two hemispheres of the brain. Also, some of them may be more dominant than others, to promote organized action.

This would mean a physical hierarchy of levels begetting consciousness, from neurons, to modules, to interconnected areas composed of modules.

Conjecture 2: The cerebellum shapes the actions of the messy brain to bring neat order and shape of execution to the intended acts started by its willful spasms. The cerebellum shapes thought motions- movements of the mind- just as with limbs and fingers, to allow for precise manipulations and basic connection and flow of action sequences. In this case the actions are on mental objects. It is known that perceiving a physical object is closely tied, circuit-wise, to manipulating it. Mental actions require some coordination, also.

14. From Association to Thinking

Thinking is mental action- studying, as though with our eyes, sets of objects that appear in the mind by sub-mind processes, and manipulating them, as though with our hands. Actions of all kinds are directed by will, which emerges from feelings. The ability to then execute a series of action objects underpins language. Language allows for more precise and useful organization and more precise and longer serialization. With this running commentary to ourselves we persuade and convince ourselves (or more precisely, some portions of self-persuade and convince other portions).

Like lower feeling, association is spontaneous and undisciplined. When motor precision is brought to bear, thinking becomes a real and elaborate act of its own, of a machine that can be relied on in complex situations. The total mind is a mix of free-run and controlled mental movements. The free-run mechanism is generally "sub-conscious" and ineffable. It injects things into our true mind, objects from some areas and words from others. It thus provides our mind's eye and our mind's ear, which recurrently see their results and respond with related, new such. Meanwhile we manipulate- move and modify- these objects and words with our motor- and speech-related faculties, and the free-run portions react to the new patterns so made.

To our minds, almost everything is either an object in space that we manipulate, an emotion, or an utterance. These concepts are generalized to represent nearly everything we think about. This presents the opportunity to construct a "physics of mind," meaning not the underlying physics in this case, but a correct theory of operation of this macroscopic world of nonextended things, in other words, metapsychology.

Consciousness Architecture Layer 3: Mind

Once we have sentience- qualia- the stage is set to begin to actually build minds. Layers 3 and 4 are mind architecture. The internals of 3 and 4 are being well-advanced by neuroscience. In this section we will make reference to brain parts and begin our tour of our theory (STHC) for Layers 3 and 4.

Cognition happens when rather than just link an informational and a pain/pleasure input to a (possibly reinforced) action to navigate in and/or manipulate the physical world, internal "objects" crafted of these sensations become themselves things to navigate to or around or, especially, to manipulate. In other words, cognition represents the birth of abstraction.

At the top of Layer 2, qualia that have been automatically generated appear from below and may be used for manipulations on the above side. There is also now a what-it-is-like to *do* something on the above side, and the demands for action are sent down for their execution by unconscious inter-cooperative acts.

Behind the forehead, what is known and what is wanted collide to result in action decisions and resulting awareness of those already made.

14.1 Animals and Vertebrates

Earth and the other planets formed 4,500 Mya (million years ago). Within the first billion years the oceans formed and then life within them (most likely at deep sea vents). At 900 Mya evolved the choanoflagellates, the last single-cell precursor to animals, and at 665 mya the first primitive, amorphous animals, at first without neurons but soon these appeared also, enabling simple reflexes and rhythms. At 585 Mya bilateral symmetry appeared.

Then, at 539 Mya, began the Cambrian explosion, lasting roughly 20 million years, and yielding all the animal phyla alive today including the craniates (animals having skulls) and the subset of those known as the vertebrates (animals also having a vertebral column replacing the notochord), at 518 Mya. All living craniates- 70,000 species- are also vertebrates.

The simplest vertebrates are the hagfish (which have no jaws). Hagfish (and lampreys)- fully functional vertebrates- have no myelinated neurons. These were followed by jawed fish (445 Mya), amphibians (350 Mya), reptiles (310 Mya), and mammals (200 Mya). Myelinization appears in jawed fish. 6-layered cerebral cortex, known as neocortex or isocortex, is unique to mammals.

All vertebrate brains consist of, from tail end to head end

Medulla	of brainstem
Pons	of brainstem
Mesencephalon	(midbrain) of brainstem
Cerebellum	
Dienchephalon	(hypothalamus and thalamus)
Telenchephalon	(cerebral hemispheres)

The medulla of the vertebrate brainstem deals with nonconscious functions of heart rate, breathing, blood pressure, and vomiting, and the pons deals with nonconscious functions of respiration.

The midbrain of the brainstem contains

- The periacqueductal gray, which contains various neurons involved in the pain desensitization pathway.

- The substantia nigra pars compacta, which is a concentration of neurons in the ventral portion of the midbrain that uses dopamine as its neurotransmitter and is involved in both motor function and emotion. Its dysfunction is implicated in Parkinson's disease.
- The reticular formation, which is involved in the pain desensitization pathway, and is involved in the arousal of cortex into consciousness, and contains the locus coeruleus, which is involved in intensive alertness modulation as well as autonomic reflexes.
- The ventral tegmental area. A dopaminergic nucleus, known as group A10 cells, is located close to the midline on the floor of the midbrain.

14.2 . Animal Mind

Layer 3 is a large layer, containing almost everything about building a real, working brain from the tiniest beginnings of one. In Layer 3 we first assemble a mind and then much improve it.

While we might have protected pockets in the “messy”, hot brain for smaller scale quantum assemblies, at this scale, expecting unity of consciousness from brain-size entangled quantum states is a tall order. A different mechanism must exhibit itself (e.g., fields) at least by the time we enter the scale of Layer 3.

This layer is where intelligence- the general ability to solve problems, appears.

This is the primary domain of neuroscience, plus animal psychology.

Lower mammals, e.g., rodents, have a smell-dominant perceptive world. Higher mammals, e.g., apes, have a perceptive world dominated by visual objects. The eyes of the latter feature full-color vision and a fovea with high-resolution. We can see this orientation every time we put abstract objects "off to the side" while considering something else (perhaps after getting a good "grasp" of them). These abilities serve well for the coming of language abilities.

The brain is built with successive systems of hardwired behaviors, habits, feeling, and computation, with new brain tissue enabling increasingly higher repertoires for behavioral choices.

A maximal Layer 3 brain can do essentially everything that can an adult chimp or 2- year-old human.

14.3. Hypothesis Contained in STHC

Hypothesis STHC-1: It is by way of feeling and information processing that aware and intelligent minds have developed that can solve problems and set goals.

14.4. Hardwired Actions in Nuclei and Ganglia

Older parts of the evolved brain and spinal cord provide a repertoire of recognizers/categorizers and motor patterns that supply mental functions with some of their mechanical needs.

14.5. Drive

A drive itself is the hardwired tendency to specific actions given specific stimuli. (We have developed a drive to take what we mentally want.) The activation of drives throws off feelings. Drives proper are earlyish developments and do not represent cognition.

14.6. Activation from the Brainstem

The brainstem (in particular the reticular formation), taking many inputs, issues diffuse stimulation into the cortex, which otherwise passively falls into low neural activity and unconsciousness. The activation process itself is mechanical and unrepresentative of either feeling or cognition.

14.7. Reinforcement Memory of Early Animals

Animals are the multicellular organisms that move. Movement, even blindly, was advantageous over being completely static, and animals that moved in useful ways survived and reproduced. Extremely helpful, next, is sensation, to trigger movements. Even single-celled predecessors have this. Everything described so far relied on evolution for this "learning," which occurs across generations.

The next big innovation was animals that could learn, by making stimulus-response links variable and providing multiple paths. This learning is what we call reinforcement learning. Reinforcement memory links stimuli, responses, and outcomes. It does not have to be centralized, it can and did develop in ad hoc fashion and location originally.

Later in evolution this memory proves insufficient because learning comes from averages of several learning events, which allows too many errors as some evolving animals face more difficult, dangerous (e.g., daylight) lifestyles featuring increased risk of predation and volatile resources for their high energy needs.

14.8. Pain (Dysphoria) in Anterior Insular Cortex (and Anterior Cingulate Cortex)

Centralized arrangements such as the vertebrate brain allow for the same circuitry to be used for learning of many different things. The same pain-pleasure-memory circuits can be applied to multiple problems as at the same evolutionary time the separate informational circuits come to connect and converge and fuse information into single elaborate objects and qualities, eventually on an internal global stage of world experience.

It appears that neurons rich in neural processes that evoke fundamental pain are rich in the Anterior Cingulate Cortex (Brodmann areas 24, 32, and 33). It appears that what actually occurs in the ACC is higher level emotionality or reflection than pure pain (Ambron, 2022). The ACC is agranular cortex, older than the granular areas. It is part of the Papez circuit.

Positionally and connectionally, the ACC lies at the interface of emotional cortex and cognitive cortex. The ACC appears to be involved in learning the association of actions and punishment (or reward) (Rolls, 2016).

Closely connected to the ACC is the insula; the anterior insular cortex is considered emotional cortex. The anterior insula appears to be where pain happens (Ambron, 2022). Tellingly, it is active for pain of psychological origin as well as that from physical damage. The anterior insula, too, is agranular, older cortex (the posterior insula is granular).

Existence of the pain areas allow for both selective storage of emotionally bound content and regeneration of the emotions as a consequence of recall.

14.9. Pleasure (Euphoria) in Insular Cortex, Agranular Orbital Prefrontal Cortex (and Anterior Cingulate Cortex)

"Hedonic hotspots" have been identified in subcompartments within the

nucleus accumbens shell, in the rostrodorsal quadrant of the medial shell. An element of the ventral striatum of the basal nuclei. These neurons receive dopamine and project to the ventral pallidum and other areas. The basal nuclei have existed since the earliest vertebrates.

posterior part of the ventral pallidum, which is an element of the basal nuclei. It projects to the thalamus.

parabrachial nucleus, in the dorsolateral pons of the brainstem; this has only been demonstrated to respond to benzodiazepine receptor agonists (e.g., Valium). It allows normally pleasurable stimuli to be pleasurable when it is inhibited (Kringelbach, 2010), which suggests the pleasure is happening somewhere higher.

orbitofrontal cortex (OFC), anterior portion (Kringelbach, 2010)

insular cortex; including posterior (Kringelbach, 2010)

All the hotspots are linked and activation of one hotspot results in the recruitment of the others, and inhibition of one hotspot results in the blunting of the effects of activating another hotspot. This tends to impede knowledge of where the pleasure is actually occurring.

Numerous studies have shown that the orbitofrontal cortex is important in the reward system (Kringelbach, 2010). For connectional reasons, (Carmichael, 1994) and (Passingham R. , 2021) include the anterior insular cortex in orbital prefrontal cortex along with the traditional Brodmann 11, 13 and 14. Caudal 13 and 14 and the anterior insula together make the agranular orbital PF cortex; the rest of OFC is granular and more recent and has been seen to be involved in much higher-level aspects such as associating perceived object identity, attributes, or classifications with their degrees of value. Lateral parts of the anterior OFC are connected to the ACC.

On balance, the evidence points to the insular cortex, the agranular orbital prefrontal cortex and perhaps the anterior cingulate cortex as the location of our pleasure experience, with the last perhaps instead more reflective (subject-oriented) than directly experiential.

As with pain, the capacity for pleasure allows for both selective storage of emotionally bound content and regeneration of the emotions as a consequence of recall.

An interesting aspect of pleasure is that, unlike pain, there are no bodily receptors for it. It is always attached to stimuli that are themselves generic in nature. This highlights the fundamental difference between data and valence.

14.10. Basic Awareness

Basic awareness is feeling.

14.11. Global Long-Term Memory and Recall Courtesy Hippocampal Formation

The hippocampal circuitry, which connects widely and in the same fashion across the cortex, enables global, episodic memory. Items linked need only have simultaneity (or close relationship to those that do).

This tissue has existed since the early vertebrates and was originally used for navigational purposes, representing place and sequence.

As there evolutionarily becomes much more to remember, the hippocampal complex retains centrality. During sleep, medium term potentiations of the day result in consolidation into long term memory as the primed areas are cycled to strengthen synapses. Dreams are a side effect.

14.12. Associative Training Throughout the Cortex

Most of our cortex is “associative” cortex, but there are different modes of actual effect.

Basic subcomponents of heard language and seen scenes form sets of recognizable primitives to which all the later incoming should be attached. This is done by repetition, not requiring feeling, and forms a repertoire of fixed recognizers. The activated recognizers along with feeling are inputs to dynamic selection circuits, resulting both in real-time flows and retention of the most salient events, some of which may be committed to long term storage during accesses during sleep.

Added to the basic recognize-select two-stage system (which is also used in electronic logic chips) is a refinement whereby the selection circuit, which has identified its relative favorite but may not have reached threshold to fire that output, is subjected to a periodic “gamma” wave of stimulation every 25ms. This wave passes recurrently from lower-processing to successive higher-processing regions, assuring each will fire and do so in hierarchical sequence and close-together times in ripples, reducing fade of inputs when needed for calculation and synchronizing the progression of analysis by processing level across the brain. When back and front regions are well synchronized as indicated by EEG, the subject is fully awake and can discern what “makes sense” from what does not, unlike in REM sleep. Gamma wave equivalents have been modeled successfully in ANNs by the author.

Spatial locations in the cortex are used to map to both real space and body part locations, finely, and broad category spaces such as vision and hearing, coarsely.

Note that auditory cortex is cortex that is attached to the ear, and the arbitrary experience evoked when an individual hears noises becomes associated, becomes what hearing is like. Hearingness qualia do not populate virgin temporal cortex. When entirely new organs are added to us, there will be a what it is like (Nagle, 1974) for them as well.

The generality of cortex promoted its expansion and local tuning.

14.13. Attention via the Thalamus and Corticocortical Backprojections

Attention allows the same cortical circuits to analyze or act on different objects at different times. This serves both economy and generality. Attention proper is a mechanical process. In mammals, attention is tightly coupled to feelings of awareness, such that impactful experience will command attention and things attended to have much fuller feeling responses. This marriage is so tight that they seem as though a single process.

The thalamus is mapped with one-to-one regions of itself and neocortex (not older cortex) reciprocally connected and features lateral inhibition resulting in concentration on certain inputs by masking the others on the way to the cortex. This serves as a preprocessor to the neocortex, making the inputs to it more sparse (Rolls, 2016). The recurrent loops may support short term memory.

Higher regions of cortex back-project to next-lower ones in service of top-down attention (Rolls, 2016).

Other nuclei of the thalamus participate in global regulation of cortical activity (Kandel, 2021).

14.14. Biased-Competition Memory of Early Mammals via the Neocortex

Neocortex appeared with mammals, and from early on included agranular prefrontal cortex. The older, reptilian allocortex (such as the hippocampus) learns quickly and is better for current conditions; the neocortex learns more slowly and stores more specialized representations and is therefore better suited for the long-term statistical realities of the world (Murray, 2017).

With the ability to have new representations comes the problem of how to handle when the different representations come into conflict as to identifying things or selecting action. Agranular cortex is well-suited to resolving these conflicts and making a choice and serves in this role as granular cortex develops and expands.

14.15. Short Term Memory, The Workspace of Mind

Short term memory is implemented as neural signaling loops. As other parts of the cortex expand or appear during evolution and higher-level concepts are introduced, more loops for short term memory are introduced to sustain dynamic access to them, expanding the prefrontal areas, and all the levels are retained across evolution and cycle simultaneously (Rolls, 2016). Short term memory greatly increased in higher animals.

The loops are made from all excitatory cortical neurons. This is inherently unstable but short term memory is extremely useful and thus valued in evolution. The loops are stabilized by the influence of inhibitory neurons. Epilepsy occurs when this balance fails.

14.16. Intention

Intention follows want and is the innate drive to action to a want. The introduction of generalized pain and pleasure centers allow for multitudinous things to become wants or things to be avoided.

14.17. Subjectivity

"Subjectivity" refers to two entirely different things, though we often use them together. The first is informational-cognitive- which is to perceive or consider things relative to oneself (or to other things).

The other refers to "subjective experience," which is elementally a happening of feeling.

These can be (and are) mixed such that one can experience the act of analyzing something from one's own perspective, or note that since a matter affects one, one can't help but have an opinion that is "subjective," which generally means emotionally tinged.

14.18. Complex Emotions

Complex emotions are formed when feeling is mixed with the cognitive (data). This is where the common supposed "raw feels" of qualia often arise, such as "the taste of a wine." Unless we believe that every such quale anyone will ever have is an irreducible experience in the universe (which is unlikely), that quale is not elemental.

Personality is defined by the responses made to the various painful and pleasurable experiences the animal has (Rolls, 2016).

The Anterior Cingulate Cortex mentioned earlier has been cited as the interface between emotion and cognition (Allman, 2006).

14.19. Mindful Self

A psychological self emerges which at bottom is essentially one strong cloud of feeling reacting to others. The borders of self-move constantly.

14.20. Manual Foraging Memory of Early Primates via New Premotor and Posterior Parietal Cortical Areas

This memory enables us to learn correctly to reach for objects in cluttered environments (Murray, 2017). It evolved to aid a new lifestyle in the fine branches of trees, with better vision and increased use of the hands for both foraging and feeding.

We reach "on autopilot" without needing to be consciousness of individual movements to the object. We can do this even if there is motion such that we must change trajectory in real time.

With this fundamental ability completely mastered, most action became literal or abstracted grabbing and manipulating objects (or vocalizing).

14.21. Feature Memory of Monkeys via Additional Posterior Parietal and Temporal Cortex

With monkeys came the fovea and trichromacy and further visual processing in the dorsal ("What") and ventral ("Where") visual streams, providing Feature Memory, enabled by additions to posterior parietal and temporal cortex (Murray, 2017). These provide our "object reality."

14.22. Goal Memory of Monkeys via New Granular Prefrontal Areas

The goal system reduced the number of errors in learning by generating goals based on single, goal-related events, new attributes and metrics provided by feature memory, and *abstract* strategies that applied solutions to familiar problems to novel ones (Murray, 2017). This was an important benefit for monkeys because of long life, dependence on the unreliable produce of angiosperm trees, and the need to forage over long distances in daylight and thus also be on the ground and exposed to predators. Mistakes were dangerous.

The goal system supports top-down attention, one-trial learning, the application of previous solutions to new problems, and arbitrary categorization.

Whereas the existing hippocampal memory system learns about what happened at a particular time and place, the new granular prefrontal areas learn about what happens in relation to behavioral goals. With the appearance of humans, these will come to work together.

14.23. Flowering of Intelligence, the Ability to Solve Problems

The mammalian model of evaluative summary feelings steering all the decision points in cognitive processes coupled with the cognitive objectification of everything into manipulable units is a highly scalable model, a leap allowing for rapid quantitative expansion of cortical areas to allow for more short-term working memory, more associative storage of relationships and tendencies, finer judgements, and increased generality, reuse, and abstraction. All of this is well-promoted by the already extant large and long-term storage capability.

Primate-level problem solving ability appears evolutionarily with granular prefrontal cortex (first appearing with the manual foraging system). The Brodmann Areas in this cortex continue to enlarge and refine over evolutionary time. We see the problem-solving capabilities of primates advance as the newer brain areas appear (Passingham R. E., 2012).

14.23.1. Problem Solving 1: Volitional Action Selection Based on Outcomes

The Medial Prefrontal Cortex enables action selection based on outcomes.

The Medial Prefrontal Cortex exists in monkeys. It consists of agranular (no layer IV) and granular cortex. The agranular portion consists of Anterior Cingulate Cortex and prelimbic cortex (Brodmann 24 and 32), and the infralimbic cortex (Brodmann 25). All mammals have homologs of these.

The granular portion consists of the medial part of Brodmann 9 and all of Brodmann 10- polar prefrontal cortex. Only primates have these areas.

14.23.2. Problem Solving 2: Evaluating Resources and Object Selection Based on Outcomes

The Orbital Prefrontal Cortex enables object selection based on outcomes.

Orbital prefrontal cortex is also known as OFC. It consists of Brodmann 11, 13, 14, (and parts of Brodmann 12, 47, and 10).

Part of areas 13 and 14 are agranular.

14.23.3. Problem Solving 3: Searching for Objects and Goals

The Caudal Prefrontal Cortex (Brodmann 8) enables searching for goals. The Caudal Prefrontal Cortex searches for and orients attention toward valuable goals, based predominantly on vision.

14.23.4. Problem Solving 4: Planning Sequences and Generating Goals Based on Recent Events

This is the point in evolution where the fovea of the eye appears and creates a central high-resolution field of vision surrounded by a fuzzier periphery. It is also the point where trichromatic vision appears, enhancing discrimination ability.

The Dorsal Prefrontal Cortex (Brodmann 46) enables generating goals based on recent events.

14.23.5. Problem Solving 5: Associating Objects and Generating Goals Based on Visual and Auditory Context

Ventral Prefrontal Cortex (Brodmann 47, 45) enables generating goals based on visual and auditory context. It generates the goal (concrete or abstract object, location, or action) that is appropriate to current context, desired outcome, and current needs.

14.23.6. Problem Solving Integration: Abstract Rules, Attentional Performance, and Generating Goals Based on Current Contexts and Events via Prefrontal Cortex as a Whole

The granular prefrontal cortex made its appearance with primates, to find and evaluate objects in a visually complex environment in the trees.

The prefrontal cortex has connections to sensory cortex, hippocampus, amygdala, premotor cortical areas (receiving goals of movements), brainstem oculomotor nuclei, claustrum, basal nuclei, thalamus, midbrain, and cerebellum. The five cortical regions described in the five prior sections are also extensively interconnected, with abilities of newer regions building on earlier ones.

Three brain information processing hierarchies, one each for context and for goals and for outcomes, feed into the prefrontal cortex. The prefrontal cortex maps contexts to goals and predicts outcomes.

14.24. Self-Consciousness

As the brain evolved to have more and more capacity to sense its internals, it became possible first to respond cognitively and affectively to its own thoughts and feelings, and then to recognize as a totality its personal thoughts and feelings, which together with inter-associated responses to bodily sensations formed in sum an experienceable object of self. The similarities between other objects acting in the world and the self-acting in the world ultimately added the understanding that the self was one of many world actors.

Self-consciousness has very little to do with consciousness in the sense that it is a high capability succeeding a long line of definitely conscious ancestors. The self is just one more object of which to be aware.

An awareness of self-precedes language, precedes humanhood. Several species (but only those) reveal an awareness of self by how they react to a mirror placed in their vicinity (Gallup, 1970). These species are great apes, an occasional elephant, rays, dolphins, orcas, magpies, and, remarkably, one species of fish (the cleaner wrasse). Monkeys fail the test; they are not self-aware.

(*Continued on Part III*)

How Brain Makes Mind: The Principles of Operation (Part III)

Karl Sipfle*

Abstract

We present the principles of operation by which a brain makes a mind, at all scales necessary to cover the whole problem. We inventory the necessary capabilities for a mind. We divide conscious mind into four layers of increasing elaboration. For the principles of operations of the lower two layers, we provide the STFC theory. For the upper two, we provide the STHC theory. We survey the evolutionary progression from first twinge of experience to human capacities. We explain the types of memory and problem-solving we carry and by what structures they are made to happen. We compare to prior works and review the philosophical implications and stance. All of this is done with minimal incoming assumptions, and those made are declared.

Part III of this four-part article includes: 15. What It is Like to Be a Human; 16. Consciousness Architecture Layer 4:

*Correspondence author: Karl Sipfle, Independent Researcher (also working independently at NASA GSFC).
E-mail: ksipfle@umich.edu

Language-Enabled Mind; 17. Evolution; and 18. Conclusions.

Keywords: Consciousness, physics, evolution, feeling, qualia, mind, cognitive, affective, sensation, memory, learning, attention, perception, recognition, decision-making, problem-solving, coordination, self, symbol manipulation, language.

15. What It is Like to Be a Human

As higher consciousness emerged there came the experience and recognition of time, as a consequence of the differences it sees between its world at times t and t0 (where t and t0 keep increasing). Effectively, consciousness moves along time. At a low level there are linked stimulations of conscious events, and the propagation times and stimulus persistence of the brain. These cause the continuous flow of high consciousness time, enabling our "now."

Some opine that consciousness *causes* time (Deutsch, 1998), in the sense of the experience of a physical dimension that would otherwise just be another physical dimension. One might say that time is the progression toward increased entropy, or one might say that time is the *experience* of progression toward increased entropy; this contrast is just a matter of definition. The idea that the consciousness of beings such as ourselves specifically and literally causes physical time, however, we reject. There is insufficient evidence and plausibility to accept that possibility. It remains possible that the sentonic field and time have a very close relationship.

Time might be a sequence of quantum collapse events, but space might, also.

Humans are action-oriented. The brain exists to decide what to do. The frontal lobe is the motor lobe. Impulses for action are suppressed and analyzed and selected and organized for submission to execution areas. Unlike other animals that go to sleep whenever they can, humans are curious and exploratory and chatty and get bored, stir-crazy, and lonesome. This keeps us active and learning.

The human brain features high-level, complex connections: understanding is pleasurable, novelty is pleasurable. These make curiosity and enhance creativity, itself a mix of compliance and variance.

Things bubble up into human consciousness from "subconscious" areas. Only some parts of the brain are organized to be part of the human-conscious. It is the unity of feeling and the unity of the cognitive and the fusion into a greater unity of these two (plus language) that form the experience of what we call our minds. For the base feeling this fusion is done in and by the sentonic field. For the cognitive, this is done by networking of neurons. The final fusion is done by the connection of cognitive (information-processing) neurons with feeling neurons, which connects cognition to feeling. This creates both complex deep emotions with a cognitive component, and rich thought embroidered with, encouraged by, discouraged by, and steered by emotion. The two things that make all this possible are the information processing and communication ability of all neurons, and the (strong) connection to the sentonic field of some neurons.

To “make sense” is a pleasurable condition related to high discovered recognition and correlation, and low ambiguity, as seen in strong standing signals against a relatively quiet background of alternatives. It brings comfort and is sought. Confusion and unsureness are negative, and we seek to eliminate or avoid these. The neural circuits for this are imaginable.

The human mind loves order in all things. It has grown through evolution to where it is possible for it to see a whole big picture (and even to describe that to another), wherefrom flows its great power.

Organized mind appears when awake, as opposed to when dreaming. One possibility is that the mind would be organized when we sleep were it not for interference caused by signals that are generated (or removed) during sleep (Sipfle K. , A Neurobiological Theory to Explain the Lack of Rationality in Dreams, 2021). Another possibility is that the operation of perceptual regions close to the senses, which are inoperative during sleep, ordinarily organize the flow through the rest of the brain, to restrict what is thought to what “makes sense.” Both factors could well be significantly contributing. A third possibility is that prefrontal centers that impose a critical evaluation upon earlier cortex are disconnected during sleep; this is supported by EEG studies (Sipfle K. , A Neurobiological Theory to Explain the Lack of Rationality in Dreams, 2021).

We note also that *emotions* are quite in their normal form and understandable even during sleep; it is the cognitive, which is intrinsically of structured nature and of high continuous relatedness to many patterns we know, that loses good form and becomes disjointed.

A related possibility is that our simulation centers- our imaginations- are only constructed to run coherently for seconds at a time, because that's what is needed. In sleep we also don't playback our memories verbatim in long sequences; memory centers are executing a consolidation process and bubbling and burping as a result when they are active at all. Thus, disconnected splotches of memories start up short sequences of reconstructions and predictive simulations with no sensory perception or memory circuits activation to police for order.

Human-level consciousness occurs from adding to lower animals

a) Full language, with its ability to precisely manipulate symbols and abstract objects

b) A strong sense of what "makes sense"

c) Much increased short-term working memory and richer interconnections

(b) is what is not acting as much when we are dreaming. We may say that a scene or utterance is "incomprehensible," which means it does not comprehensively hang together (make connections) that result mostly in recognitions of things and relationships deeply honed over the years, rather than confusion resulting from a lack of this. Things make sense when there is high recognition and sharp inhibition, connected to feeling. In sleep the contrast is lost, and things bleed over into each other.

Humans are so advanced that they are aware of concepts as high as feeling confused and feeling sure they have "got it" (which are indeed feelings and may even be misplaced ones). Knowledge is not feeling. But feeling convinced of knowing something is a feeling (and generally a positive one). "I feel that X" means that X is coupled with a feeling of some certainty.

The mechanics of a working mind run on Recognition and Value. Direct pain and pleasure served as the first source of Value, allowing bootstrapping. Now, many local inter-neuronal circuits have likely generalized Value informationally, and no longer feel pain and pleasure, there being no evolutionary need to maintain them. At the micro level of large networks, the original pain and pleasure are too crude to serve- at least, globalized pain and pleasure are too crude.

Mouse consciousness is surely emotional and very action-oriented. It lives in a world of smells and touch and motor movement. In contrast, a grown chimpanzee has the intelligence of a 2.5-year-old human, and probably his emotional life as well. Such a human is no longer an infant but a toddler, with rudimentary language ability. Human consciousness is visual and analytic. Humans inhibit action until selected and initiated after analysis of the world.

Returning to time, consciousness has *duration*, necessarily. The actual "remembered present," as Edelman called it (Edelman, 1993), gives a more *global* (across space) and *constant* (across time) existence to an object than the millisecond/millimeter world of neural-neighborhood-scale operations.

This very short-term memory is very useful to sophisticated cognition, a sort of object cohesion and permanence enabler. Coupled with the connections that compare between different parts of this time snippet and those that react to it and supply it with emotion, to a first approximation this *is* low-level, higher-animal consciousness. It is the now in which all thought and feeling occurs, about three seconds long.

Consciousness has a smidgeon of usefulness even without net emotional affect. Change and novelty and distinct form cause this impact (which on repetition fades, both as it happens and from (nonmotor) memory). But consciousness without net emotional affect, which supplies positive and negative values, is very limited in what it can do.

The emotions we feel affix value to things, including objects to study and courses of action to choose. In principle in another universe another mechanism could have been used to supply value but feeling is what evolution discovered in our universe that could be used.

And why did evolution use this means? What is helpful about feeling, is that it has its own intrinsic value- positive or negative. That means evolution did not have to manage to use valueless connections in a consistent way to emulate the same thing, or invent centers as assigned positive and negative places that must be connected to. Consistently labeled positive and negative things were available in nature to happen upon, and then use again and again. This made it easier, and thus more likely, and thus the way, that summary value is implemented in the brain.

We don’t call it thinking unless we can feel it happening. There are also underlying processes to our thinking which we can’t see (e.g., associating). Crossing the boundary is what happens when things “pop into my mind.” We are able to deliberately reflect to encourage these unconscious processes to happen, and to deliberately turn down our attention and fixations to allow broader and freer flow of proto-thoughts and metathoughts.

We can feel the thinking process partly because cogitating kicks feeling parts (rather like feeling limb motion), and probably because of the changes in blood vessels (which we can see in imaging).

It is probably even important to our thought processes that we can feel our thoughts and where we are paying attention, and probably there is specific sensation of that because the sense-ability has utility.

General “awareness” may include arousal (which increases intensity of attention behavior) more than emotional response, which gives value to things. A high level of arousal can be pleasurable or stressful; it is an intensifier.

Our emotions are different from an awareness of things. We can just feel sad or fearful, *then* start thinking or noticing or inventing supposedly why. Sad can be purely dysphoric, without knowing “why,” which is cognitive (in contrast, the more complex *dread* is also dysphoric, but *refers* to something).

People use the word “feeling” to refer both to emotions and to intuitional judgements. This is natural, because our processing circuitry includes a bed of elaborate, general recognition circuits effectively much-interwoven with feeling circuits so as to develop finely detailed value processing of things under consideration, resulting in a net high-level score, and the distinction between emotion and cool judgement can be difficult for us to tease apart.

It is possible both to have a “feeling” (observationally) of a large object in front of your face without net caring, and also possible to feel positive or negative for no apparent reason. As to the former- impact- it is the shapeless glob of feeling that is doing the reacting to factual stimuli. You still hear emotional white noise even though it doesn't settle on organized tones. The white noise activity is impact. Everything noticed makes "you" feel something- even if only a unified agitation.

Concentrating of pain or pleasure is relatively unusual in the universe (rather like the very big molecular machinery of life being unusual). Thus, it is a specific achievement of evolution that is useful. Pain and pleasure allow advanced minds. By the Anthropic Principle, that is why it's here.

Physical consciousness was exploited in vertebrates just as the laws of mechanics and chemistry were. It serves the functions of learning and assigning general goodness and badness to large and complicated and long-lasting experiences and plans, and in bonding social groups.

Higher consciousness is like multicellularity, something that emerged as a complex system incorporating many points of occurrence of base consciousness, allowing novel survival niches by supporting highly elaborate, flexible, and effective structures.

While our consciousness is observable and organized, this does not mean that simple consciousness does not happen all around us. The organization of fundamental consciousness into high-level consciousness was done with our brains by evolution (or preceding mind to whom we are the artificial intelligence). As one of the available forces, fundamental consciousness probably has been exploited elsewhere, and elsewhere there will be pain and pleasure (and therefore perhaps good and evil).

Is consciousness "just" an epiphenomenon? Yes, in the sense that it was not necessary that we actually feel, only that the same values be attached to things as feelings allow. No, in that the values consciousness provides form a key part of what our minds are and how they work.

Value is needed for how we think. It is what steers the cognitive centers to make decisions. The cognitive centers can recognize things and recall candidate action sequences, but it is the emotional input that determines the selections-the "voluntary" behavior. Volition equals emotion.

The nature of our human-level consciousness is that we react emotionally to (and perceive) our thinking, and we cognitively notice our feeling, and we talk to ourselves about all of this in a chorus of stream of consciousness (James, 1890). Our words then evoke feelings and more thoughts and around it goes. We talk to ourselves like a friend (or enemy), and our real friends' words are in there, incorporated.

I feel. What is I? An assemblage of interlinked cognitive and feeling events. I am intellectually aware that I am feeling, and I can feel and have feelings about my thinking. The human mind is everything it is because it contains not only thought and feeling but can think about feeling and feel about thinking, and do so under the organizing and evocative presence of language. Higher consciousness is the interaction of knowing and feeling, and each of each other. The deep mutual interplay of feeling and thought, and of these with language, comprise the mind.

16. Consciousness Architecture Layer 4: Language-Enabled Mind

There are two major elements added to make a human level mind beyond just scale. These require new brain tissue. One is a refined sense of what "makes sense" (that which is lost when dreaming). One might call this our reality detector. The other is language.

Language organizes all the little fits of proto thought, and one's own assembled phrases stimulate those same fitful centers to recognize and remember and churn out new proto thoughts. Language crisply organizes associations and spatial, motor, and intuitive thought. This is done using the new language centers of our brain (notably Brodmann 22, 44, and 45). The appearance ("hearing") of the sentences to our non-language-centric brain centers evokes in them the unstructured thought that is then organized into language again. This cycle, along with emotion, is the fundamental stuff of our highest-level consciousness (mind), as we talk to ourselves about our thoughts and feelings and react to what we have heard from ourselves.

We have coined many words for various kinds of feeling (and thinking), which attest to the richness and complexity of human-level minds.

16.1. Human Evolution of the Prefrontal Cortex

While 99% of human DNA is the same as a chimpanzee's, 80% of the proteins differ between the two species (Passingham R. , 2021), which diverged 7 Mya. By about 2

Mya humans (Homo Erectus) had evolved, left Africa, and made it to China. In another wave, by about 0.6 Mya archaic Homo Sapiens appeared and made it to Germany. Modern Homo Sapiens appeared by 0.3 Mya in Africa, 0.2 Mya in Greece.

The frontal cortex- particularly the granular prefrontal cortex- and medial parietal cortex, was expanded in Homo Sapiens.

During the evolution of the anthropoids (monkeys through humans), there was an expansion of association cortex within each lobe, and not just within the frontal lobe. The human prefrontal cortex, though, expanded more than the middle and inferior temporal cortex, the premotor, and the motor cortex. This means there is more human prefrontal cortex for re-representing visual information from the ventral visual stream and for generating goals as opposed to just directing the actions that achieve them (Passingham R. , 2021).

16.2. Goal and Feature Memory of Humans via Parietal-Prefrontal Network, Temporal Area, and Prefrontal Adaptations

Parietal-Prefrontal capability advanced from representing relational metrics to supporting generalized relational reasoning. Temporal areas advanced from representing the signs of resources to supporting semantic generalizations. Prefrontal cortex adapted its top-down biasing function to memory retrieval. (Murray, 2017).

16.3. Social-Subjective Memory of Humans via New Granular Prefrontal Cortex

New and higher levels of hierarchy emerged courtesy additional prefrontal areas. Representations of self and others developed in these areas. These came to influence existing medial and lateral networks (Murray, 2017).

The medial network came to support perspective-taking, recognition of situational contexts, mental simulation of past and future events, and knowledge of self and others from participatory experience.

The lateral network came to represent social goals, norms, and concepts, categories, groups and roles of individuals, and generalizations about self and others.

16.4. Explicit Memory of Humans

Explicit memory emerges from interactions between the forms of memory, each having evolved at different times. The hippocampal memory system learns about what happened at a particular time and place while granular prefrontal areas learn about what happens in relation to behavioral goals. With the appearance of humans, these came to work together (Murray, 2017).

High order, human-specific representations of self-contribute to both the perception of participating in attended events, and the perception of knowing attended facts. These self-representations become a part of conjunctive representations that are explicit memories. When individuals retrieve memories containing this part they re-experience the participating in events and knowing of facts (Murray, 2017).

16.5. Language-Enabled Mind

On Earth, only humans operate at this level. Other species have not achieved true grammatical language. Universal capacity for grammar is literally wired into humans (Pinker, 2009).

As we have noted, observation of the experiential parts of the mind by the verbal parts results in the spontaneous forming of sentences that crystallize precise statements of observation. This evokes responses of meaning in the experiential parts of the mind, forming our loop of realize-and-comment that operates continuously in our waking hours.

These mental abilities enable long-chain serial thought and precision and abstract languages such as mathematical notation.

(None of this capability has anything to do with Goedel. Recognizing mathematical truths is a combination of learned discernment skill and natural machinations of our evolved brains that live in a naturally patterned world. Also, many people cannot make these recognitions so it cannot be fundamental to the human mind.)

Language is long, sequential chains of symbols, especially with the advent of writing, which provides external memory much as the real world of objects is itself persistent.

16.6. Hypothesis Contained in STHC

Hypothesis STHC-2: Cyclically, language organizes thought and stimulates new thought to organize as language centers interact with earlier non-language centers (Sipfle K. , 2018).

16.7. Problem Solving 6: Reasoning, Imagination, and Planning and Generating Goals from Instructions and Imagination via Enlarged Polar Prefrontal Cortex of Humans

Polar prefrontal cortex is Brodmann area 10, located right behind the forehead. Brodmann 10 is now the largest Brodmann area.

Through human-level general intelligence exploiting human capacity for instruction, imitation, and mental trial and error in internal re-representations of the world, humans are able to reduce errors in action choices- which are dangerous in the wild- to nearly zero. The imaginative abilities, even of actions planned but not executed, that emerged to reduce errors in specialized situations, created the internal world that provides modern insight (Passingham R. E., 2012).

16.8. Problem Solving 7: Language via Areas 44 and 45 and Culture and Social Rules via Medial and Orbital Prefrontal Cortex and Amygdala

Functional brain imaging studies and meta-analyses thereof have revealed the following knowledge (Passingham R. , 2021)

There is cerebral dominance in the left for speech, but also for performing skilled actions other than speech, regardless of hand used. The same precision manipulation mechanisms are used.

The prefrontal area Brodmann 44 is active when words and sentences are being assembled (but not when speaking them). Left areas 44 and 45B support generating verbs appropriate for nouns (that is, verbs that make sense for nouns) , with 44 generating sentences by hierarchical rules. 45A supports learning association pairs and names associated with things (tagging with words). Names can evoke visual imagery (elsewhere in the cortex) of the named objects.

Left 44, 45B, and 45A plus the middle frontal gyrus are critical for the semantic system.

Observing demonstrations actives 44 and 45B. Verbal instructions can be acted on immediately without first learning. Instruction activates prefrontal cortex linked to cortical areas specific to the requested tasks.

The interconnected medial and orbital prefrontal cortex and amygdala are activated in anticipating punishment, feeling empathy, and feeling guilt.

17.Evolution

We have presented an architecture, which itself stands for study irrespective of the specific explanation as to how it is "implemented" by Nature. But as evidence of its truth, it must be a plausible destination of evolution. Here we summarize that journey. We can recapitulate the evolutionary development sequence because the needs and difficulties are obvious, and because we have the animals around us as evidence.

First needed would be basic stimulus/response. This could be viewed as the degenerate case of the cognitive. Then would come totally dispassionate ability for slightly complicated perception and actions. In other words, calculating ability- pure, passive cognitive- would appear first, and still exists in primitive animals.

Feeling provides for the first operant (reinforcement) learning opportunity, and certainly for more complicated organisms. It offers a broad whatever-you're-doing-is-bad (or good) signal for learning, and it provides the second needed, broadly distributed system- Value- which determines what action impulses are let through, the first system being Activation. Third is Attention, targeting and focusing the cognitive portions. Even before well-developed learning appears, feeling allows for generalized real-time changes in decisions in creatures with some neural complexity, and serves Attention.

The wholesale march into feeling began with the evolutionary outgrowth of the allocortex- the olfactory system and the hippocampus. The paleocortical piriform cortex, a.k.a. posterior orbitofrontal cortex (posterior OFC), is heavily correlated with the cingulate gyrus and the septal area, which is part of the basal forebrain. The OFC mediates the expectation of reward/punishment in response to stimuli. The paleocortical olfactory tubercle plays a role in transmitting positive signals to reward sensors.

Once all of these lower mechanisms were in place, the neocortex burst forth into high growth. Warmer/colder is a powerful concept (complementing recognition and association).

The older and deeper parts of mammalian cortex form the connections to pain and pleasure that guide the cognitive circuits (literally) above. With this basic plan in place, the stage was set to scale up- with the same basic brain architecture- to more and more, and more refined, cognitive mass. Summary good/bad/pain/pleasure/warmer/colder is key to advanced brains and simplifies genetics in that action selection circuits can be built to pursue summary "good" and avoid "bad," while other independent ones connect perceptions and relationships to those values, rather than trying to wire together myriad situational aspects to individual acts of plans.

As the ability to manipulate objects became more abstract, true language first emerged and then became not just a communication feature but integral to human thought; communication itself became internalized and reflective. Metalanguage (grammar, new notations, compound constructed words, and descriptive theories) emerged and itself became introspectively manipulable.

Pain and pleasure have been discovered and harvested by evolution. (There are likely other worlds in which pain and pleasure don't exist to organisms, even complicated ones.) Pain and pleasure might not actually be the same force. And there might be animals that feel pain but no pleasure. Use of one may have come first in bio-evolution.

Intensity of experience increases by way of more spots in the field, and so indirectly, by more related brain interconnections. Harvesting fundamental consciousness by evolution would mean "crafting" the shapes and/or construction and constituents of neurons, and possibly sets of neurons.

Feeling continues to guide cognition at all levels. This allows three-pound brains to handle situations usually well enough that would be impossible to handle with combinatoric perfection, and provides an edge to conquer, in evolution, competitors and challenging niches. With bacteria and others optimized for the easier (in terms of complexity) niches, we populate the harder ones, requiring large-scale sophistication.

If not already on Earth, comets or asteroids could have provided batches of particles rich in sentonic charge. They could have come from anywhere or anyone. They are the seeds of advanced minds.

18.Conclusions

We have presented an Architecture (FLA) and compliant theory for every major constituent of mind from nothing but the existence of this universe up through human minds.

The STFC metatheory does not presume that human-level consciousness is everywhere. It does assert that primitive fundamental consciousness is distributed throughout the universe and the field carrying it is everywhere. The universe does not understand things or feel as a whole in a unitary fashion. It does contain such "flecks."

The STHC theory of animal-level consciousness assumes that it is probably the result of activity of networks of physical neurons that calculate and harness feeling.

The nature of our human-level consciousness is that we react emotionally to (and perceive) our thinking, and we cognitively notice our feeling, and we talk to ourselves about all of this in a chorus of stream of consciousness. Our words then evoke feelings and more thoughts.

On the way to our solution, we have laid down some postulates and presented hypotheses, to wit:

Postulate 1: Physicalism is necessary and information - and computation-based theories cannot provide for the source of conscious experience.

Postulate 2: Consciousness functions within the same basic rules (including mathematics) as all other phenomena that also exist in Nature.

Postulate 3: Something basic is missing from, and must be added to, our physics.

STFC Hypothesis 1: Fundamental feeling is freestanding and requires no separate feeler. At most there is a physical interaction between two fundamental elements.

STFC Hypothesis 2: There is a physical process that is fundamental feeling.

STFC Hypothesis 3: The fundamental feelings are pain and pleasure.

STCF Hypothesis 4: There exist pain and pleasure in minimal discrete bits independent of any brain, at the sub-molecular scale.

STFC Hypothesis 5: Some biological brains have evolved to organize and exploit pain and pleasure.

STFC Hypothesis 6: The fusion of feeling across the brain occurs in a force field.

STHC Hypothesis 1: It is by way of feeling and information processing that aware and intelligent minds have developed that can solve problems and set goals.

STHC Hypothesis 2: Cyclically, language organizes thought and stimulates new thought to organize as language centers interact with earlier non-language centers.

(*Continued on Part IV*)

How Brain Makes Mind: The Principles of Operation (Part IV)

Karl Sipfle*

Abstract

We present the principles of operation by which a brain makes a mind, at all scales necessary to cover the whole problem. We inventory the necessary capabilities for a mind. We divide conscious mind into four layers of increasing elaboration. For the principles of operations of the lower two layers, we provide the STFC theory. For the upper two, we provide the STHC theory. We survey the evolutionary progression from first twinge of experience to human capacities. We explain the types of memory and problem-solving we carry and by what structures they are made to happen. We compare to prior works and review the philosophical implications and stance. All of this is done with minimal incoming assumptions, and those made are declared.

*Correspondence author: Karl Sipfle, Independent Researcher (also working independently at NASA GSFC).
E-mail: ksipfle@umich.edu

Part VI of this four-part article includes: 19. Summary Discussion; 20. Comparison to Other Theories; 21. Philosophical Position; and References.

Keywords: Consciousness, physics, evolution, feeling, qualia, mind, cognitive, affective, sensation, memory, learning, attention, perception, recognition, decision-making, problem-solving, coordination, self, symbol manipulation, language.

19. Summary Discussion

The fact that it is difficult to understand does not make consciousness any less a natural phenomenon than electricity or nuclear fusion (Chalmers, Forward: Addressing the Hard Problem of Consciousness, 2017). The Standard Model incorporates everything we completely understand so far, and in the long run never fails to accommodate extension and revision to cover new discoveries.

Being quite distinct from the other phenomena, a new force and thus field, charge, and boson are scientifically called for. The new force interacts, by some chain of interactions, bidirectionally with the electromagnetic force, which is responsible for nearly all operations in the living brain.

Evolution has learned to harness this "new" force, which has existed latently for eons. Learning how to interact with it directly will allow the creation of feeling machines and the continuity of our developing selves for centuries.

Electromagnetism is already heavily used in making up the fine informational computation and transmission patterning in the brain. The new force provides something "out of band" from that to connect with to provide the emotional value.

Clouds of odd and differing shapes in a mist of consciousness "vapor" provide both valent experience as their essence and distinguishable larger-scale feelings that serve as meta-symbols (tags) around which clear and distinct mental objects can be constructed and then manipulated, like snowflakes around a mote of dust.

This paper has dealt with the necessity of explaining the *beginnings* of our minds in two steps:

1. The fundamental capacity in this universe to feel anything, at its most basic and indivisible.
2. The combining of this into the lowest level tier of experience that we can find introspectively from within our minds. This is the bootstrap from Layer 1 to Layer 2.

1 and 2 in place can then be assumed and provide the conceptual atoms and form the complete platform- the interface- upon which minds can be built, in stages of cognitive capability.

With 1 and 2 in place there is sentience, and the mind from there becomes simply the result of increasingly fancy computation with connection to value, and self-observation, focus, and manipulation. We call that Layer 3. Work on Layer 3 is proceeding apace and is not the hardest consciousness problem today.

The platform of Layers 1 and 2 could be simulated with effectively exact fidelity, even without knowing the precise physical interaction used by our universe, allowing artificial "consciousness" that does not actually feel but is just as efficacious (Sipfle K. , Distinct Kinds of Consciousness, 2019). (The model also allows real man-made consciousness to be inserted when that is mastered.)

We must have stuff to move around in order to represent and manipulate information. Computationally, it doesn't matter what the stuff is. Any stuff that exists in self-local little pieces can be arranged.

We need a certain specific thing to be feeling- specific stuff. Feeling by itself on the local level has nothing to do with information, nothing to do with arrangement.

The fundamental magic of our minds is the marriage of information processing with feeling, and in particular, valent feeling. The information is the arrangement of stuff. The essential stuff is feeling.

We see:

- We need stuff that feels. Not just any stuff.
- We need stuff to arrange and rearrange. Any readily arrangeable stuff will do.
- The stuff can be entirely the same, that is arranged, or there can be a somewhat more complicated arrangement between the two. Either way, as a practical matter, just a little way up it is the same stuff.
- Feeling goes way down. Valence goes way down.

- Feeling does two things- fuse/join and thus generalize and simplify into grosser states, and provide on the same many scales a sense and score of good/bad.

- Subjective feeling is an epiphenomenon just in the sense that any natural mechanism that had the utility just described would do. In our universe pain and pleasure are available for the task and have been harnessed.

- Electromagnetic processes are both coupled with and have a high degree of separateness from feeling. The separateness provides new utility; feeling does a "processing" by which the electromagnetic processes are enhanced, as proven by evolution weaving feeling in.

Only three basic things are needed to explain the mind

- Physics-level understanding of feeling.

- Explanation of path from physics level to bottom-most mind (qualia in a unified place).

- Cognitive processing mechanisms, ultimately including language. These are coming into view via neuroscience from microanatomy and various scanning and probing and psychological experiments.

These things are reflected in the corresponding architecture of the mind:

4. Language and formal reasoning.

3. Cognitive processing, pre-language.

----

2. Building from physics to bottom of mind leading to integration of cognitive with feeling and formation of bottom mind experiences and actions.

1. Physics: Manipulable objects to represent information, feeling action to summarize and evaluate.

Where there has been much confusion is in 1 and 2, and this is what tells us what consciousness *is*- in other words the difference between a simulated consciousness and an actual consciousness. We have not yet detected the base constituents. As with many other advances in earlier times with thin data, we are driven to our conclusions by reason, and this time we know how Nature tends to work.

Meanings not only feel like something in an active mind, they are made entirely out of feelings, in that even the informational content is patterns- patterns of feelings. This brings together the informational and the emotional aspects of knowing.

While information/cognitive is the patterns and independent of the stuff, pain and pleasure are stuff of a particular kind and independent of information patterns. The two become associated early and often and form the fundamental fabric of what is our consciousness. As this dance proceeds higher, we know what we feel and that we feel, and we feel that we know and what it is like to know.

The existence and operation of real material in our world demands both atoms and all the complexities of the separate-yet-joined-into-wholes way in which these atoms participate, and this looks to be almost exactly what is happening with our most basic experiences, as we should expect. A cell *is* made of atoms, even if upon hearing about atoms you still don't see how that could explain life. That just means there are more steps of explanation and structure than are describable simplistically. What is required is not just to avoid oversimplified explanations but also oversimplified presumptions of what we are starting with.

The true brute facts occur at the bottom and amount to nothing other than the rules from which the universe is made out of its energy. There has to be *something* standing in that place. These will include the Physical Correlates of Consciousness that underly and explain the Neural Correlates of Consciousness.

A fundamental consciousness operating at the scale of quantum physics underlies all conscious experience. At this level there are only particles affecting each other, consciousness is a freestanding physical event, and there are no observers present.

Typical qualia occur at a higher level and are elaborations incorporating "psybits" of fundamental consciousness. That qualia are themselves composites is shown by introspection and by the realization that, otherwise, the universe would need thousands of different essential flavors of consciousness, an explanation which is both suspiciously bloated and very unlike the rest of the known universe. Additionally, varietal compositions (based on forces) are how all the rest of the known universe has organized into higher structures.

Since psybits affect each other at the physics level, so do the "molecules" that make up qualia.

Components of different qualia have distinguishable feels because the molecules of psybits have different exact makeup and internal arrangement, made inevitable by variations in the physical brain.

If we gradually anesthetize a brain, human consciousness eventually disappears, at a point before almost all neural firing has stopped. Likely, the longer and thinner links between locations of high self-supporting connections break first, leaving islands of independently conscious chunks of brain that no longer result in a unitary mind. As activity continues to fall, these islands of activity with their internal tighter loops will also collapse. At the most extreme nonzero case, sputters of fundamental consciousness will be evoked, but these will be both too weak and too simple to amount to any anything mental. They will be simply more noise in the universe. But they may someday be detectable!

A simple experience to us involves many neurons supporting each other's firing, each (conscious) one of which probably causes many individual physics-level events. A simple thing to the mind is large on the particle scale. Even a weak feeling to a person is rather rich in absolute terms.

What we call our consciousness is that which is accessible in the unified mind. Outlier bits of consciousness will exist in other parts that are not connected in, in our own brains, other peoples' brains, and probably in the whole universe. The boundaries of the unified mind are probably dynamic as things enter and leave our "consciousness."

Researchers who see consciousness as fundamental in the universe without recognizing that it is assemblies that make qualia in minds, and psycho-philosophers who think their red is a natural primitive, are talking past each other in trying to make the connection between their worlds. Physics is not high enough and psychology is not low enough for that. Even introspectively, we can see that red is not an irreducible sensation, and neither is "redness" a moving particle.

We see that the large question as to what a whole human mind is made of can be well-split into two pieces, which are basically the domain of the Hard Problem, and a lot of information computation and storage. Therefore, the solution may be sought by different specialties and teams. As to applications, understanding of the feeling half will have limited importance for the creation of effective artificial (and stoic) "minds," but be of central importance to extending our own conscious lives forward.

The key link in understanding the whole of consciousness is the explanation from self-less physics to the much higher level of qualia, at the bottom edge of the introspectable mind. Key clues might be found in pain and pleasure, which appear to have a much simpler and more direct link "upward" than the typical qualia.

20. Comparison to Other Theories

There are two ways in which other theories of consciousness differ from the STFC+STHC presented in this paper: sweep across scale and specifics within their scope. None of the below theorists cover the full multiscale sweep of consciousness. Below shows their coverage.

IV Edelman James

III	Crick	Damasio Tononi	Eccles	Edelman	James	Koch	
II		Tononi?				Koch?	
I	Pribram	Tegmark	Eccles Tononi			Koch	Penrose

20.1. Pribram

Pribram made many contributions to the base of neuroscience upon which all theories now stand and which live in Layer 3.

The unique contribution of Pribram's Holonomic Brain Theory is the position that some processing occurs in dynamic electric fields that surround dendritic trees, and further that these fields can link neurons (ephaptically). In his original form it relies upon quantum effects; further development is being done (Poznansky, 2022). Evidence continues to appear for ephaptic signaling of some kind (Sipfle K. , Support for the Sentonic Theory of Fundamental Consciousness (STFC), 2019) (Anastassiou CA, 2011) (Chiang, 2018) (Francis JT, 2003), and it's required for STFC for the link between electromagnetic operations and the feeling force. Thus STFC bears some rough similarity to parts of HBT but differs in others, such as STFC requiring a new force and not demanding quantum behavior.

20.2. Penrose

Penrose's OR theory (Penrose, The Emperor's New Mind, 1989) (Penrose & Hameroff, Consciousness in the Universe,

2017) addresses only Layer 1 and misses the distinction between fundamental and higher consciousness. However, Penrose believes mind is made from microscopic bits of consciousness that are physics acts, which is a shared view with STFC+STHC. A notable predecessor was John Eccles.

20.3. Edelman

Edelman (Edelman, 1993) did not address Layer 1 or 2 but did discuss a language Layer 4 as layered upon an animal Layer 3. He also discussed primary repertoires of neural pattern recognizers being recruited for secondary repertoires. Additionally he discussed the "remembered present" of the several seconds of immediate conscious experience in our mind, where lives what others called the global workspace. He also expected this current short term memory lives in recurrent cortical loops. All of these beliefs are compatible with STHC.

20.4. Damasio

Damasio does not describe Layer 1 or Layer 2 but does vaguely acknowledge a need for something in their place (Damasio A. , 2021) (Damasio A. R., 2003). His work is centered on Later 3 and highlights the criticality of feeling, which he describes accurately for a Layer 3 perspective. STHC mostly agrees.

20.5. Koch

Koch's work has been mechanistic, calculational, and information-oriented. He has worked much with Crick and then Tononi. With Crick he postulated that oscillations and synchronization are necessary bases of consciousness. Today, where STFC agrees with Koch is in his belief in some kind of panpsychism (discussed later). However, he is a supporter of IIT, with which STFC disagrees. Koch is a proponent of consciousness as a result of complexity (one of the Primary Pitfalls of (Sipfle K. , The Primary Pitfalls on the Road to Understanding Consciousness, 2021)), with which STFC also disagrees. Koch has not elucidated an understanding of the primacy of minimal feeling vs. the higher consciousness that is built from it. Yet, through questionable reasoning, he does state that consciousness ultimately cannot be derived from anything else, a tenet of STFC.

20.6. Tononi

Tononi's elaborate IIT (Tononi, 2012) relies on a position that information begets feeling. Information is a capturing of pattern of something else, information is always about something. Information itself contains nothing and is nothing, certainly nothing physical. There is no analytical argument that can be made about such a fundamental assumption, the argument is simply definitional and ontological (and epistemological), which is that the assumption is contradictory to all foundations and makes no sense. In other words it is flawed philosophically before one gets to the (theoretical) science. This author's theory is in disagreement. All this also means, however, that the theory potentially could be rehabilitated by replacing the assumption with one recognizing the need for a patterned energy of some kind.

20.7. Tegmark

Tegmark is another primary proponent of the notion that complexity yields consciousness (Primary Pitfall #11 (Sipfle K. , The Primary Pitfalls on the Road to Understanding Consciousness, 2021)) and sees perception, self-awareness and subjectivity free of a feeling (or any other ultimate) basis. STFC disagrees.

20.8. Rolls

Rolls has issued many detailed thoughts regarding consciousness (Rolls, 2016). He opines that "consciousness may be the state that arises in a system that can think about (or reflect on) its own (or other peoples') thoughts, that is, in a system capable of second- or higher-order thoughts." This is about higher-order consciousness, and also not about feeling. Rolls's writing on consciousness is largely about Layer 4. It does not conflict with STHC.

He seems to be beginning to notice the importance of feeling with "The system that is having syntactic thoughts about its own syntactic thoughts (higher-order syntactic thoughts) would have to have its symbols grounded in the real world for it to feel like something to be having higher-order thoughts." However, his immediate explanation mentions only data-oriented descriptions. He also says "The gene-specified rewards or goals for action that are the bases for emotional and motivational states play the role for these states of grounding them in the real world" which sounds promising but the actual argument dances around feeling and avoids invoking it. He then points out that he has still not explained qualia and proceeds to attempt so in an entirely informational way.

He says "the present proposal has as its foundation the type of computation that is being performed, and suggest that it is a property of higher-order syntactic thought system used for correcting multistep plans with its representations grounded in the world that it would feel like something for a system to be doing this type of processing" and "representations needed to be grounded in the world for a system with higher-order thoughts to be conscious," understanding a need for "grounding" but not quite where it comes from.

He seems to have finally pounced on the truth with "I suggest that one type of meaning of representations in the brain is provided by their reward or punishment value," yet concludes "the causal role of consciousness is that the information processing that is related to consciousness …can play a causal role in producing our behavior" and then then "it does feel like something when this processing is taking place," which sounds very epiphenomenal.

Rolls says "my view is that consciousness can be implemented in neural networks…, but that the neural networks would have to implement the type of higher-order linguistic processing described…, and also would need to be grounded in the world."

Finally, he says "Consciousness, and especially phenomenal consciousness (why it should feel like something) is a 'hard' problem in philosophy. It is not clear that it is amenable to a solution using neuroscience." This is correct. What must be added is physics. This provides the feeling that cannot be understood through neuroscience alone. This provides the "grounding" that both gives meaning to symbols of any sort and explains why mentation feels like something.

Further down in our stack Rolls proposes that binding is provided by feature-combination neurons rather than stimulus-dependent synchrony as forwarded by Crick and Koch. This does not conflict with our theory.

21. Philosophical Position

Descartes's view of consciousness (DesCartes, 1644) was that it was something other than natural. He introduced dualism. We disagree.

Fifty years later, Leibniz introduced monads (Leibniz, 1720), repudiating dualism. He proposed everything actual is made of a few classes of minimal objects. His objects were created monads, monads that are souls with perception and memory, and monads that are rational souls. Consciousness was thus assumed, in fact, central. We agree with monism and the fundamentality of feeling.

A hundred years ahead of his time, Whitehead realized matter was not part of the ultimate nature of things. He also grasped the centrality of relation, in anticipation of today's work by physicist Carlo Rovelli (Rovelli, 2017). He believed in the universe all entities experience, although he had a very broad definition of "experience." His views led to David Ray Griffin coining the term "panexperientialism" to mean a view that all entities experience to distinguish it from panpsychism, the idea that all matter has consciousness. The position of STFC is that some universal entities, not necessarily matter, have fundamental consciousness also commonly known as "subjective experience." If "matter" is taken to mean at the gross level of biology, then certain matter (e.g., brain tissue) will contain the necessary elements for consciousness while other matter may not. This may be called panexperiential materialism (Poznanski & Brandas, Panexperiential materialism: A physical exploration of qualitativeness in the brain, 2020) and is consistent with STFC.

Searle says that consciousness is a real subjective experience and is caused by the physical processes of the brain (biological naturalism). STFC agrees with this, with the proviso that "the physical processes of the brain" most critically mean the fundamental-physics-level processes that the brain invokes. Simply "naturalism" may be a better term.

Chalmers, holding some beliefs matching Searles's, formulated the hard problem of consciousness, which (though smaller) much like the FLA and STFC+STHC provides useful, categorizing structure for further thought and work. Chalmers identified "panprotopsychism," which holds that higher-order phenomenal properties (such as qualia) are a logical consequence of protophenomenal properties, at least in principle. This closely mirrors STFC, with our clarifications that the "protophenomena" are already feeling, not things that will result in feeling at a higher level, and that they are ubiquitous but not associated with everything.

Protophenomenal properties are by definition the constituent parts of consciousness. Chalmers has argued that the view faces difficulty in dealing with the combination problem. We answered that problem in Layer 2.

In summary our philosophical view might be called Essential Experience Ubiquity, or simply Ubiquitous Feeling.

Conflict of Interest Statement: The author holds no relationship presenting a conflict of interest for this work.

References

Allman, J. M. (2006). The Anterior Cingulate Cortex: The Evolution of an Interface between Emotion and Cognition. *Annals of the New York Academy of Sciences, 935*(1), 107-117.

Ambron, R. (2022). *The Brain and Pain.* New York: Columbia University Press.

Anastassiou CA, P. R. (2011). Ephaptic coupling of cortical neuron. *Nat Neurosci, 217-223.*

Beuhl, S. (2015). Towards a "canonical" agranular cortical microcircuit. *Frontiers in Neuroanatomy*(14 Jan).

Carmichael, S. T. (1994). Architectonic subdivision of the orbital and medial prefrontal cortex in the macaque monkey. *Journal of Comparative Neurology*(346).

Chalmers, D. (1995). Facing up to the problem of consciousness. *Journal of Consciousness Studies*, 2: 200-19.

Chalmers, D. (2016). The Combination Problem for Panpsychism. In *Panpsychism.* Oxford University Press.

Chalmers, D. (2017). Forward: Addressing the Hard Problem of Consciousness. In *Biophysics of Consciousness.* Singapore: World Scientific Publishing.

Chiang, C.-C. (2018). Slow periodic activity in the longitudinal hippocampal slice can selfpropagate non-synaptically by a mechanism consistent with ephaptic coupling. *The Journal of Physiology*.

Crick, F. (1995). *The Astonishing Hypothesis: The Scientific Search for the Soul.* New York: Simon & Schuster.

Damasio, A. (2021). *Feeling & Knowing: Making Minds Conscious.* New York: Penguin Random House.

Damasio, A. R. (2003). *Looking for Spinoza.* London: Heinemann.

DesCartes, R. (1644). *Principia philosophiae.* Amsterdam:: apud Ludovicum Elzevirium.

Deutsch, D. (1998). *The Fabric of Reality.* New York: Penguin.

Edelman, G. (1993). *Bright Air, Brilliant Fire.* New York: Basic Books.

Fain, G. (2014). *Molecular and Cellular Physiology of Neurons 2nd Ed.* Cambridge, MA: Harvard University Press.

Francis JT, G. B. (2003). Sensitivity of neurons to weak electric fields. *Neurosci,*, 7255-7261.

Gallup, G. J. (1970). Chimpanzees: Self recognition. *Science, 167*(3914), 86-87.

Gamez, D. (2018). *Human and Machine Consciousness.* Cambridge: Open Book Publishers.

Harris, A. (2020). Consciousness Isn't Self-Centered. *Nautilus, 2.*

Hebb, D. O. (1949). *The Organization of Behavior.* New York: John Wiley & Sons.

Hemmings, H. C. (2019). Towards a Comprehensive Understanding of Anesthetic Mechanisms of Action. *Trends in Pharmacological Sciences, 40*(7).

Heuvel, M. P. (2011). Rich-club organization of the human connectome. *Journal of Neuroscience, 31*(44), 15775-86.

James, W. (1890). *The Principles of Psychology.* Cambridge, MA: Harvard University Press.

Kandel, E. (2021). *Principles of Neural Science, Sixth Edition.* New York: McGraw-Hill.

Kringelbach, M. L. (2010). *Pleasures of the Brain.* Oxford: Oxford University Press.

Leibniz, G. (1720). *Lehrsätze über die Monadologie.* Christian Wolff.

Mountinho, S. (2021). Edison Was Right: Waking up right after drifting off to sleep can boost creativity. *science.org.*

Murray, E. (2017). *The Evolution of Memory Systems.* Oxford: Oxford University Press.

Nagle, T. (1974). What is it like to be a bat? *Philosophical Review,* 435-450.

Passingham, R. (2021). *Understanding the Prefrontal Cortex.* Oxford: Oxford University Press.

Passingham, R. E. (2012). *The Neurobiology of the Prefrontal Cortex.* Oxford: Oxford University Press.

Penrose, R. (1989). *The Emperor's New Mind.* Oxford, UK: Oxford University Press.

Penrose, R., & Hameroff, S. (2017). Consciousness in the Universe. In *Consciousness and the Universe* (pp. 8-47). Cambridge, MA: Cosmology Science Publishers.

Pinker, S. (2009). *How the Mind Works.* New York: W. W. Norton.

Poznanski, R. (2019). Theorising How the Brain Encodes Consciousness Based on Negentropic Entanglement. *Journal of Integrative Neuroscience.*

Poznanski, R., & Brandas, E. (2020). Panexperiential materialism: A physical exploration of qualitativeness in the brain. In

Quantum Boundaries of Life (pp. 301-367). Cambridge, MA: Academic Press.

Poznansky, R. R. (2022). The activity of information in biomolecular systems: a fundamental explanation of holonomic brain theory. *Journal of Multiscale Neuroscience, 1*(2).

Rolls, E. (2016). *Cerebral Cortex: Principles of Operation.* Oxford: Oxford University Press.

Rovelli, C. (2017). *The Universe is Not What it Seems.* New York: Penguin Random House.

Sipfle, K. (2018). *The Nature of Consciousness.* New York: Barnes and Noble Press.

Sipfle, K. (2018). The Nature of Fundamental Consciousness (preprint). *PsyArXiv*.

Sipfle, K. (2019). Distinct Kinds of Consciousness. *RG*.

Sipfle, K. (2019). Support for the Sentonic Theory of Fundamental Consciousness (STFC). *RG*.

Sipfle, K. (2019). The Solvability of The Hard Problem of Consciousness and its Relation to Fundamental Consciousness and Human Consciousness. *ResearchGate*.

Sipfle, K. (2021). A Neurobiological Theory to Explain the Lack of Rationality in Dreams. *ResearchGate*.

Sipfle, K. (2021). The Primary Pitfalls on the Road to Understanding Consciousness. *Journal of Neurology and Neurobiology, 7*(1).

Tononi, G. (2012). *Phi: A Voyage from the Brain to the Soul.* New York: Pantheon.

van Kerkoerle, T. (2014). Alpha and gamma oscillations characterize feedback and feedforward processing in monkey visual cortex". *Proceedings of the National Academy of Sciences.*

The Neural Correlates of Feeling That Explain Consciousness

Karl Sipfle, Independently

at NASA GSFC

ksipfle@umich.edu

Written March 5, 2023

Abstract:

Our conscious minds think and feel. Thinking deals in information. Feeling is the single defining distinction between consciousness and nonconsciousness. Known brain structures and their behaviors have been proposed as the Neural Correlates of Consciousness. To fully be so they must be plausibly associated with both the informational and the affective aspects of consciousness, especially the latter. It turns out that previously precisely specified sites in the brain for NCC are largely correct though they may have to be more inclusive, and the rationalizing theory must be extended to incorporate feeling. A full understanding requires physics and the extension of existing physics; the latter is presently required anyway for other reasons. The implications lead to world-changing technologies. Extensive supporting argument is in the references.

Keywords: consciousness, feeling, neural correlates of consciousness, recurrent connections, pyramidal cells, axon, ephaptic, physical field

Argument

Our conscious minds do only two things: think and feel [1]. The interplay is rich and deep. Thinking deals in information. This part can be completely implemented with conventional computers. What makes the control systems of humans and other animals entirely different from any machines of today, be they artificial or natural such as systems within cells, is that these animals feel [2]. Feeling is the single defining distinction between consciousness and nonconsciousness.

"Feelings" such as we experience are often large and complex and come with information attached. The noninformational essence is feeling. This is what is meant by "subjective experience" and is the central matter of what Chalmers calls the Hard Problem [3].

It has been proposed to locate Neural Correlates of Consciousness. It has been proposed that these are, macroscopically, the recurrent loops between neurons thought to implement short term memory [4] [5]. We accept that these loops are indeed the machinery of short-term memory, as the information carried by neural firings is made usefully to persist in this manner.

Information is not and cannot make feeling [6] (a key fact not understood by leading theories of consciousness, e.g., [7]). Finding neural correlates of information persistence does not sufficiently demonstrate understanding of consciousness. However, it turns out that this neural correlate of informational regeneration and persistence is also in its manifestation a correlate of the same, and more, for feeling.

That is because the axonal firing of some neurons, by a process not yet understood, causes the physics behavior that is feeling [2].

Microscopically, neurons produce spikes that each can be taken in isolation to represent one bit of information. Different neuron types have output waveforms that vary from others' but that is not material to our discussion. The bit of information is transmitted on to the neurons that receive the spike.

There is a next higher level of information once we introduce time, in the form of patterns of spike trains across time. These, too, are ordinarily rather short-lived and do not provide enough to make our short-term memory.

The firing of a neuron can also produce a "psybit" (term coined by Eccles) of feeling. This is not transmitted in the same manner as is the information and is not of the same nature. It is something additional to the processes that sufficiently transmit the informational signal.

As with all other indivisible behaviors of our universe, the feeling is an occurrence in a physical field, in this case a feeling field. Opinions vary as to the identity of this hypothesized field. Top contenders for this include something entirely "new," and a scalar field that is dark energy (Sipfle K. , The Nature of Fundamental

Consciousness (preprint), 2018)- in which case most of the energy/matter of the universe is consciousness (feeling energy). Others take it to be some aspect of electromagnetism. It is almost certain to at least interact with electromagnetism, as EM explains nearly everything about the brain that we do understand. In any case essentially all such theories acknowledge that the present Standard Model of particle physics will have to be refined, as is already required to solve other major problems in physics [9] [10].

As stimulations of that consciousness field, these bits of feeling influence each other at different locations in the field, as with all other forces [11]. Thus, there is a fusion [12], explaining the unity of feeling despite its discretely located generations.

The field also affects the operation of neurons, closing a causal loop. Otherwise, feeling would have no utility and would not have evolved in us into such an elaborate, present, and much-used feature of us. This represents a second key path of communication between neurons (that is, an “ephaptic” one [13]), one that carries the valent component of our minds.

The recurrent loops that enable the regenerative persistence of short-term information also provide persistence of feeling, giving it presence across time. Therefore, the cited neural recurrent loops providing short term memory are indeed where our consciousness lives, even though there is much more to that story than was originally understood.

Not all neuron types need to excel at generating or at sensing the consciousness field stimulations. An individual recurrent loop may circulate both information and feeling or just one, and loops with different "skills" may be connected. Importantly, the originally cited loops may have to be extended in reach to include all the necessary neurons (feeling tissue).

A final elaboration is that the oppositely directed axons of recurrent connections making a loop, lying near each other, would also affect each other.

Our own experienced "now" is around 3 seconds long [14]. Eventually we should be able to trace that to the happenings described here.

Evolutionary Sequence

First, neurons were used for one-shot spike (trains) to provide real time informational processing. Any early discovery and exploitation of feeling would be primitive-fleeting bits that did not linger and align with others. For a long time, this was all there was.

There was discovered by evolution the technique of connecting neurons such that they synapsed on themselves (such as in controlling our eye position muscles [15]) or to other neurons that then return a connection to the first (possibly in multiple hops). This was done first for information processing purposes and leads to the half of our "present" that is computational.

With the emergence of these loops, as an additional effect, the utility of the consciousness (feeling) field was greatly enhanced and provided an effective summarizing out of band (back channel) communication of how things were more generally going. This is a very powerful model for building smart brains and with the basics in place brains expanded greatly in size and capability. It is not only a very useful architecture, but facilitates its continued evolution, because then genes need only specify to wire together chunks of brains in support of any actions that will produce "good" results rather than specifying topologies for individual actions and outcomes [5].

We feel because we are creatures of this universe, and feeling exists in this universe as a means of providing brains such as ours (anthropic principle) [16]. A few other means might also work in other universes.

Which Cells?

The above theory also indicates an answer to the question of which cells, of all the types, feel? The loops are made of pyramidal cells in our brains [5], therefore at least some types of pyramidals feel. Pyramidals are very old and come in many subvarieties that have been refined over the eons.

Explanation of General Anesthesia

It is known that if general anesthetic is gradually administered [17], firing of widely distributed cortical neurons gradually declines until, while neurons are still firing, loss of consciousness occurs. When neural firing rates fall below a certain point, the recurrent loops collapse, and the animal then feels nothing.

Experimental Testability

The loops of short-term memory have a physical reach of 1-3 mm, with different sets of loops having some spatial overlap [5]. Two animals scalp to scalp will well exceed that distance between those loops in their brains and the field effects would not operate as "designed."

However, two appropriate independent chunks of neural tissue featuring operating loops, such as from organoids, juxtaposed, should demonstrate the ability of one to affect the processing in the other ephaptically. Note that this tissue may well experience small amounts of pain [18] or pleasure.

Examination of how myelination or lack thereof [19] may affect the details of the consciousness field structure around axons may yield additional clues.

Technology Development

The present argument suggests the production (not merely simulation) of artificial truly conscious entities (Gamez, Human and Machine Counsciousness, 2018), or truly conscious extensions of existing entities.

The primary track entails creating a test bed of neurons exhibiting neural chunk 1 affecting processing in neural chunk 2 as described in the last section. From there development would continue in revisions toward a desired outcome with special emphasis on stripping away as much of the biology as possible, such as via transgenic organisms or organoids, to shrink the demonstrated locus of necessary action.

From the other end, artificial axons could be constructed molecularly from scratch with increasing fidelity until the same effect is observed, and then trimmed down as far as possible as well. This would result in microscopic constructions permitting huge functional scale, eventually at low cost.

Conclusion

It appears that recurrent neural loops are indeed explanatory neural correlates of consciousness even though there is much more to the picture than first assumed. Two very different kinds of regenerations are used and may be handled in varying degree by various parts of individual loops and by different, connected loops.

The chief action to establish consciousness is not just neural firing and not strictly oscillations for their own sake, but an operating recurrent neural loop executing the attendant biophysics. The micro item necessary to enable that is the twinge produced by each firing.

Further refinement of understanding by reproducing the expected effects as described and refining the means will lead to the ability both to produce new artificial minds and to permanently and meaningfully extend the lives of existing ones, such as us.

The author asserts no conflict of interest for this work.

References

[1] A. Damasio, Feeling & Knowing: Making Minds Conscious, New York: Penguin Random House, 2021.

[2] K. Sipfle, The Nature of Consciousness, New York: Barnes and Noble Press, 2018.

[3] D. Chalmers, "Facing up to the problem of consciousness," *Journal of Consciousness Studies,* pp. 2: 200-19, 1995.

[4] F. Crick, The Astonishing Hypothesis: The Scientific Search for the Soul, New York: Simon & Schuster, 1995.

[5] E. T. Rolls, Cerebral Cortex: Principles of Operation, New York: Oxford University Press, 2016.

[6] K. Sipfle, "The Primary Pitfalls on the Road to Understanding Consciousness," *Journal of Neurology and Neurobiology,* vol. 7, no. 1, 2021.

[7] G. Tononi, Phi: A Voyage from the Brain to the Soul, New York: Pantheon, 2012.

[8] K. Sipfle, "The Nature of Fundamental Consciousness (preprint)," *PsyArXiv,* 2018.

[9] R. Penrose, The Road to Reality, New York: Random House, 2004.

[10] C. Rovelli, The Universe is Not What it Seems, New York: Penguin Random House, 2017.

[11] R. Feynman, Elementary Particles and the Laws of Physics, New York: Cambridge University Press, 1987.

[12] D. Chalmers, "The Combination Problem for Panpsychism," in *Panpsychism*, Oxford University Press, 2016.

[13] G. B. Francis JT, "Sensitivity of neurons to weak electric fields," *Neurosci,,* pp. 7255-7261, 2003.

[14] G. Edelman, Bright Air, Brilliant Fire, New York: Basic Books, 1993.

[15] E. Kandel, Principles of Neural Science, Sixth Edition, New York: McGraw-Hill, 2021.

[16] R. Nozick, Philosophical Explanations, Cambridge, MA: Harvard University Press, 1981.

[17] H. C. Hemmings, "Towards a Comprehensive Understanding of Anesthetic Mechanisms of Action,"

Trends in Pharmacological Sciences, vol. 40, no. 7, 2019.

[18] R. Ambron, The Brain and Pain, New York: Columbia University Press, 2022.

[19] G. Fain, Molecular and Cellular Physiology of Neurons 2nd Ed., Cambridge, MA: Harvard University Press, 2014.

[20] D. Gamez, Human and Machine Counsciousness, Cambridge, UK: Open Book Publishers, 2018.

6. The Location of Consciousness in the Brain

The Location of Consciousness in the Brain

Karl Sipfle, Independently

at NASA GSFC

ksipfle@umich.edu

March 11, 2023

Abstract: We circumscribe the likely sites of subjective experience in the brain.

Keywords: consciousness, feeling, neocortex, agranular, pyramidal cell, axon, ion channel

Introduction

The brain does two things: think and feel, that is, the cognitive and the affective. This deep distinction runs all the way from behavior and disorder down to philosophical ontogeny [1]. On the one hand we have data computations, on the other we have the rather mysterious "subjective experience," of deeply different character [2] [3]. It is the latter that separates conscious things from nonconscious things.

Pain and pleasure are pure and clear examples of feeling and provide valuation. Experiment has revealed that these feelings are experienced in the anterior insular cortex and further appreciated in the anterior cingulate cortex [4] [5]. These are neocortical areas (but agranular [6]). They are considered prefrontal [7]. (They also contain the Von Economou neurons in humans and other animals with strong minds.)

Cortical Structure and Its Implications

The neocortex has a 6-layered structure [8].

Layer 1 (superficially) is populated mostly by connections. These are to dendritic tufts of the pyramidal cells of layers 2,3, and 5.

Layers 2 and 3 feature pyramidal cells (layer 2 has some granular cells as well).

Layer 4 contains excitatory granule cells (and some associated pyramidals). Agranular neocortex omits this layer, which appeared in evolution after both layers 2, 3 and layers 5, 6.

Layers 5 and 6 both contain large pyramidal cells.

Now we discuss the main connections [9]. Input ultimately originating at the senses enters a neocortical area at the excitatory granule cells of layer 4. Layer 4 cells mainly project into layers 2 and 3 and synapse on those pyramidals. Where granular cells are not present the input is directly to layers 2 and 3.

The pyramidals in 2 and 3 project in 3 directions. They project forward to layer 4 of the next higher area of processing. They project downward to synapse onto layer 5 pyramidals. And some project backward more diffusely to layer 1 of preceding areas, and to their layer 5 and 6 pyramidals.

Layer 5 pyramidals project to layer 6 pyramidals. Layer 5A (shallower) pyramidals also project backward diffusely to layer 1 of preceding areas, and to their layer 5 and 6 pyramidals; they also project up to L2 and L3 neurons-more on this later. Layer 5B (deeper) pyramidals also project to thalamus and basal nuclei.

Layer 6 pyramidals project weakly, to layer 4, long-distance cortical targets, and thalamus.

Forward-traveling information arrives in layer 5 primarily by the downward projections from layers 2 and 3 directly above. Layer 5 projects with imprecision to preceding areas. Layer 5B (and 6A) project to thalamus, which provides another means of modulating cortical activity.

The limbic structures hippocampus and amygdala also project into layer 1. Layer 1 and the apical dendrites terminating there appear to be more for macro energizing and priming somewhat broader subareas rather than specific data processing.

Thus we see that layers 4, 3, and 2 form the primary computational processing system and layers 5 and 6 take from them each stage's result as input and influence proceedings more generally [10].

Layers 2-4 live in a world that is data-oriented and optimized for precise analysis. It has no need to feel, although it does benefit from a net-value data input summarizing and originating from feeling.

Layers 5, 6 serve to enhance activity in portions of preceding areas they conclude to be most interesting, thereby serving attention, aiding recall by energizing those portions somewhat broadly related to what is being calculated (because the cortex uses locality to hold related things), and assists in learning [10]. A valuation of positivity (good or relevant) is what is useful here. Feeling (subjective experience) provides this (Sipfle K. , The Nature of Fundamental Consciousness (preprint), 2018). Pleasure and pain are felt in layers 5-6, by the operation of their pyramidals [12]. These cells may affect each other ephaptically (Sipfle K. , The Nature of Fundamental Consciousness (preprint), 2018) [13] [14] [15], modifying the processing in these layers. These deep cells are large and have long axons and may generate strong stimulation of the consciousness field (Sipfle K. , The Nature of Fundamental Consciousness (preprint), 2018) [15] and have a large footprint to receive them. The physical separation from layers 2-4 may be needed to insulate them ephaptically. Layers 5 and 6 feature lower resolution and smoother processing than 2-4 and have backprojections in direct horizontal fashion across the areas creating a more monolithic processing corpus.

The deep pyramidals (5/6) project to hippocampus and basal nuclei, which relate to emotion. 2/3 pyramidals do not.

Short Term Memory of Data and of Feeling

Short Term Memory is provided in the prefrontal cortex via recurrent loops between pyramidals within so-equipped areas [10]. These loops exist within both the 4-3-2 system (with reach on the order of a millimeter or two) and the 5-6 system. In their totality they provide both our conscious “now” of about 3 seconds [16] and our thinking workspace that can hold about 4 high level objects (items) for about 20-30 seconds.

Neocortex evolved a return excitatory path from layer 5A pyramidals to layer 2-3 pyramidals [10]. This creates recurrent loops between the recurrent loops within 2-3 (computational) and the recurrent loops within 5-6 (feeling), causing either being active to excite the other. The computation and the feeling within an individual area are linked. The key addition is, once again, from deep pyramidals.

Of course, there also exist between pyramidals the feedforward and feedback connections between areas which means that areas that do not directly feel pleasure and pain directly are connected with areas that do. In this case the former are the site of active ideas and concepts, the latter are the feeling.

With mutual activation the cognitive and the feeling are woven into a whole with which we are familiar.

Zeroing In

Next we ask *which part* of layer 5-6 neurons cause feeling [17].

There are only two qualitatively striking differences about the makeup of neurons vs other cells.

Ion Channels

Voltage gated Ion channels have existed since primitive cells and are nothing new [18]. However, two types of ion channels are new [19].

The voltage-gated sodium channel is an innovation (derived from the earlier voltage-gated calcium channel), and fundamentally enables the transmission of the neural spikes down the axon.

A neuron cell body has approximately 1 voltage-gated sodium channel per square micrometer, while the axon hillock and initial segment of the axon have about 100–200 voltage-gated sodium channels per square micrometer and the nodes of Ranvier along a myelinated axon have about 1000–2000 such channels per square micrometer. [20] This clustering of voltage-gated ion channels involves cytoskeletal associating proteins.

The ligand-gated channel is also new and allows the fast chemical synapses of brains. These are controlled by the binding of molecules to the channel; these molecules are called neurotransmitters.

Conceivably the channels of interest could be TRP/TRPL ion channels. We do not pursue this because for now there is less data here and they are a relatively less revolutionary development.

The electromagnetic spatiotemporal patterns of the ion currents through the channels of a neuron in operation are remarkable and may explain the link between electromagnetism and the consciousness field [21].

Axons and their Filaments

The often parallel and recurrent axons of neurons maintain a precise, small diameter and often run for long distances and contain an unusually long and unique array of filaments, also of precise structure and size, making up the cytoskeleton [19]. These include microfilaments, intermediate filaments, and microtubules.

The microfilaments and microtubules are not particularly unusual, the same as in other cells, but they (and the intermediate filaments) experience unusual electromagnetic pulses as the axon fires.

The intermediate filaments of neurons have a different makeup from those of other cells and vary between animal phyla as well and are called neurofilaments. Neurofilaments are found in vertebrate neurons in especially high concentrations in axons, where they are all aligned in parallel along the axon forming a continuously overlapping array.

Mammalian neurofilaments are heteropolymers of up to five different proteins. The five neurofilament proteins are arranged in different combinations in different nerve cell types.

Kaufman and Penrose have hypothesized that microtubules are the locus of consciousness [22].

Thus, the "greater axon" consisting of the entire length from the axon hillock to both sides of the synapse is the locus to search for the essence of subjective experience.

The means of neural response to feeling seems even more mysterious than the generation of feeling, but we know that it occurs, otherwise our brains would not make such elaborate presence and use of it. It is discussed in (Sipfle K. , The Nature of Fundamental Consciousness (preprint), 2018).

Conclusion

The distinguishing characteristic of consciousness is feeling (subjective experience), also known as P-consciousness and "what it is like." We expect this occurs at the large pyramidal neurons in layers 5-6, somewhere along the length from the axon hillock to the far side of the terminal synapse, at voltage-gated sodium ion channels or postsynaptic neurotransmitter-binding ion channels or at the axonal cytoskeleton.

The author asserts no conflict of interest for this work.

References

[1] R. Nozick, Philosophical Explanations, Cambridge, MA: Harvard University Press, 1981.

[2] D. Chalmers, "Facing up to the problem of consciousness," *Journal of Consciousness Studies,* pp. 2: 200-19, 1995.

[3] K. Sipfle, "The Primary Pitfalls on the Road to Understanding Consciousness," *Journal of Neurology and Neurobiology,* vol. 7, no. 1, 2021.

[4] R. Ambron, The Brain and Pain, New York: Columbia University Press, 2022.

[5] J. M. Allman, "The Anterior Cingulate Cortex: The Evolution of an Interface between Emotion and Cognition," *Annals of the New York Academy of Sciences,* vol. 935, no. 1, pp. 107-117, 2006.

[6] S. Beuhl, "Towards a "canonical" agranular cortical microcircuit," *Frontiers in Neuroanatomy,* no. 14 Jan, 2015.

[7] R. Passingham, Understanding the Prefrontal Cortex, Oxford: Oxford University Press, 2021.

[8] E. Kandel, Principles of Neural Science, Sixth Edition, New York: McGraw-Hill, 2021.

[9] S. M. Sherman, Functional Connections of Cortical Areas, Cambridge, MA: MIT Press, 2013.

[10] E. T. Rolls, Cerebral Cortex: Principles of Operation, New York: Oxford University Press, 2016.

[11] K. Sipfle, "The Nature of Fundamental Consciousness (preprint)," *PsyArXiv,* 2018.

[12] K. Sipfle, "The Neural Correlates of Feeling That Explain Consciousness," *ResearchGate.com,* 2023.

[13] P. R. Anastassiou CA, "Ephaptic coupling of cortical neuron," *Nat Neurosci,* Vols. 217-223, 2011.

[14] G. B. Francis JT, "Sensitivity of neurons to weak electric fields," *Neurosci,,* pp. 7255-7261, 2003.

[15] D. Chalmers, "The Combination Problem for Panpsychism," in *Panpsychism*, Oxford University Press, 2016.

[16] G. Edelman, Bright Air, Brilliant Fire, New York: Basic Books, 1993.

[17] H. C. Hemmings, "Towards a Comprehensive Understanding of Anesthetic Mechanisms of Action," *Trends in Pharmacological Sciences,* vol. 40, no. 7, 2019.

[18] B. Hille, Ion Channels of Excitable Membranes, New York: Oxford University Press, 2018.

[19] G. Fain, Molecular and Cellular Physiology of Neurons 2nd Ed., Cambridge, MA: Harvard University Press, 2014.

[20] W. M. V. W. Safronov BV, "Axonal expression of sodium channels in rat spinal neurones during postnatal development," *J. Physiol.*, vol. 514, no. 3, p. 729–34, 1999.

[21] K. Sipfle, The Nature of Consciousness, New York: Barnes and Noble Press, 2018.

[22] R. Penrose and S. Hameroff, "Consciousness in the Universe," in *Consciousness and the Universe*,

Cambridge, MA, Cosmology Science Publishers, 2017, pp. 8-47.

Part II: Reference Material

7. Distinct Kinds of “Consciousness”: Human, Artificial, and Simulated

Distinct Kinds of “Consciousness”: Human, Artificial, and Simulated

Karl Sipfle

ksipfle@umich.edus

March 27, 2019

A Reference Scheme Defining Levels of Consciousness

In preparation for what follows, we deal with the tremendous breadth of the term “consciousness,” and the vagueness, confusion, and disorganization that comes with that, by clearly defining and labelling several levels, at a granularity fine enough to be useful and coarse enough to avoid most unproven assumptions. (Sipfle, The Nature of Fundamental Consciousness, 2019)

Level 1 Consciousness consists of the smallest elements that represent experience.

Level 2 Consciousness consists of everything that must be added to level 1 to make the sentience of the simplest sentient organism.

Level 3 Consciousness contains everything above level 2 that enables the behaviors and experiences of the most complex of nonverbal organisms.

Level 4 Consciousness is that which has been attained by humans.

As it happens, this scheme also allows us to answer the question: is the universe conscious? No at Level 3 or 4. Yes at Level 1. In fact, it must be, to yield us.

The Hard Problem of Consciousness and Its Solution

The Hard Problem of Consciousness is to understand what subjective experience actually is. (Harnad, 1995)

We can see that it is not just the flow of information. There is no way for that to feel pain or pleasure (speculations to the contrary likely stem from perceived lack of other candidates). There is no way to get from an is (information) to an ought (value).

"Information" is just the patterns. Patterns are *of* something. To manifest, information must be represented by something actual. And it is in the actual medium where feeling also happens.

Information flows support the cognitive processing (regardless of what physical means supports them). Specific physical things flowing (or otherwise acting) are needed to account for pain and pleasure. (Sipfle, The Nature of Fundamental Consciousness, 2019)

We have also seen that we can't forever have an experiencer *of* something else external to it as we drill ever further down. At some point pain is just pain, happening. Something else may cause it to happen, but then it happens, *singularly and freestanding*, at the most fundamental level.

We have seen this before in the real world: a particle moves, a force exerts, a packet of light travels. All of these fundamental happenings of Nature we witness without distress. So too is feeling- fundamental feeling. When an electric charge is near another, each experiences a pull (or push), and moves. Some other such fundamental occurrence feels.

Subjective feeling at its base happens at Level 1, a piece pf physics we do not yet understand but compatible with that which we do. As with cognitive operations, the subjective experiences of our brain (qualia to us) are built from elaborations of this basic feature of the universe. (Sipfle, The Nature of Consciousness, 2018) This explanation is both sufficient and necessary to explain the subjective. It also tells us that any so-called mind that does not harness these mechanisms does not feel.

Simulation

A simulation of consciousness is not consciousness any more than a simulation of an airplane is an airplane. It is not real consciousness, either natural or synthetic. This is because consciousness is not just flows of information.

Conscious minds *feel*. They don't only think they feel or notice that they feel, they feel.

Any confusion about whether something other than represented information is necessary for feelings in our conscious minds is swept away in considering pain and pleasure instead of, say, "red," which is a more factual and neutral thing. In fact, in the case of pleasure there are not even pleasure receptor neurons; pleasure is an additional step and different response to many different inputs that at the outset often have informational functions of their own that are not related to pleasure.

Simulations do not feel pain or pleasure. Simulations are not consciousness.

Synthesis

A suitable cognitive process in concert with *actual* feeling is consciousness, whether of natural or synthetic origin, exactly as aspirin is aspirin whether extracted from nature or synthesized.

Why These Distinctions Matter

There are at least three reasons why drawing these distinctions is more than just a pleasant exercise.

1. Doing so helps us to understand what our minds actually are, as well as the "minds" of current and future robots.

2. Machines that do not experience actual distress or joy do not require rights of their own and can be exploited fully.
3. Extension in time of our own minds is not satisfied by fusion with simulated consciousness. True synthetic consciousness is required for survival.

Machines

Machines do not require true Synthetic Consciousness to be every bit as useful as if they did. They simply need a cold stand-in that produces numbers representing equivalent positivity and negativity to steer the cognitive circuits.

These can very successfully feature "Machine Consciousness," which is typically a (contradictory or extending, depending on how one looks at it) pairing and redefinition of words, as in "my reality."

But it is a better term than Synthetic Consciousness or Artificial Consciousness, both of which imply the re-creation- not simulation- of something already known. In this case, "Machine" becomes a qualifying, clarifying word.

People

Since it is expected that at bottom real experience happens with the same simple particles, the extension of the life of our minds by synthetic means is 100% correct and legitimate, just as taking in food molecules to substitute for others is completely legitimate and leaves us the same person; one electron or proton is just as good as another, and one literally cannot tell the difference, even at the atomic level. This is completely different from some kind of "upload," in which case one would have a photocopy or even brother of oneself, but the self definitely dies.

Feeling Machines, Artificial People

All of the foregoing also makes clear that with enough knowledge and by doing the right things we probably could create *machines* (not software) that *actually feel*. The question that immediately arises is: why would we want to do such a thing? It certainly raises massive ethical questions.

One view would be as a logical extension of genetic engineering. Why hack indirectly at length on the existing method of creating people, when you can just create one as you want "it" from a clean sheet of paper? "The only thing about a man that is a man is his mind. Everything else you can find in a pig or a horse," noted Archibald MacLeish. A slightly more expansive view is that it might be our natural and honorable evolution in the large, being born of stardust and fundamental consciousness, to give birth to the next big step from the same things.

Pragmatics

Words mean things. It is likely that the term Artificial Consciousness will be much seen in the wild, not because it is accurate but because it is tantalizing (if the term "Synthetic Consciousness" achieves prominence for the same thing, that will be an even worse and misleading name). Indeed, much of what is called Artificial Intelligence is a series of tricks that have little to do with what we call "intelligence" in normal discourse, which amounts to general problem-solving ability. Real Artificial Consciousness as a pursuit would involve such things as inventing brain prosthetics to combat the decay of our feeling brains.

The more accurate "Simulated Consciousness" not only has less pizazz but suggests you have the second-rate imitation and likely won't survive the marketing instincts from which terms that stick often come. "Machine Consciousness" is the best of the discussed terms that is likely to see adoption for the "mind" in the "brain" of a robot. Thus, use of the term "Machine Consciousness" is to be preferred.

Conclusions

To talk about consciousness and lead to its understanding, it is very helpful to get specific and divide this very nebulous and inclusive term into named and clearly delineated constituents.

Humanity has solved the Hard Problem in theory and steps should now be devised to refine the gross theory into specific candidates and devise means for experimental test.

Consciousness without true feeling is not really consciousness at all (unless we simply redefine "consciousness" by fiat). Simulation of feeling is not feeling.

Machine Consciousness, which is a contradiction in terms or a redefinition, but a convenient and popular one, does not require true feeling.

Extension of our minds into the long future does require true consciousness. This is *actual* "synthetic consciousness." It is to be created by humans and real and the same as the natural form.

References

Harnad, S. (1995). Why and how we are not zombies. *Journal of Consciousness Studies*, 164-167.

Sipfle, K. (2018). *The Nature of Consciousness.* New York: Barnes & Noble Press.

Sipfle, K. (2019). The Nature of Fundamental Consciousness. *RG*.

The Primary Pitfalls on the Road to Understanding Consciousness

Karl Sipfle

Independently, at NASA GSFC

ksipfle@umich.edu

December 19, 2020

Abstract

Consciousness and the human mind have been pondered for centuries [1] [2]. Today the most fundamental, likely outlines of the explanation of consciousness are known, and from this landscape we are well-positioned to fill in the basket of specific theories that will include the correct one. However, many errors in thinking and assumptions have accumulated, and even gained fashion, that obscure our progress. It is time to eliminate these, which we do herein by enumerating the complementary true assertions. Following that we mention four principles for choosing direction for further work. We conclude with a framework for further progress.

Keywords

Consciousness, brain, mind, feeling, qualia, subjective experience, physics, cognitive, affective

Contents

Definitions

Feeling: "Subjective experience." That which makes the Hard Problem [3] hard.

Raw feeling: Basal, indivisible feeling.

Qualia: The rawest feelings discernible to a mind.

Consciousness: Feeling and, typically, information processing affecting each other in organized fashion to make action decisions in the interests of a species.

Levels of Consciousness: A layered progression of Fundamental Consciousness, Sentience, Animal Consciousness, Human Consciousness [4].

Fundamental Consciousness (FC): The smallest element of what distinguishes consciousness; the actual smallest bit of feeling. Pure pain or pure pleasure. Somewhat similar concepts have been called pre-consciousness, protoconsciousness [5], and microfeels [6].

Higher-Level Consciousness (HLC): Everything above FC.

Mind: An island of Animal or Human Consciousness. On Earth, requires and belongs to at most one brain.

Cognitive: related to thought and/or information processing and not to feeling (though in practice feelings typically accompany the cognitive).

Affective: related to feeling, not cognitive.

The Traditional Raw Feelings Aren't

Qualia are discovered by the introspection of a mind. These are the experiences that specific minds report as simplest units of experience that they are able to discern. A tragic leap has been made from the simplest *to a mind*, to the simplest *of all*- that is, to the universe that has forged the mind. This is a massive and incorrect assumption [4].

The traditional qualia such as the taste of a wine are not fundamental feels. The experience of a headache is not a fundamental feel. However, Pain (and likewise Pleasure), unqualified and unelaborated pure dysphoria, are fundamentals. They are "micro-qualia," and (in sufficient synchronized numbers to emerge in a mind) are also qualia.

How do we know pain and pleasure are fundamental? Because you cannot make “hot,” valent feeling out of “cold,” neutral components, no matter how assembled. As with atoms, neutral and distinguishable assemblies can be made from the valent, but not vice versa. Even if a neutral item can be split into a positive and negative pair, this would a) have to be on the scale of base units to yield significant net valence, and b) yield what is fundamental beneath conscious minds- valent feeling.

Information is capture of pattern and can encode valence but is not itself valent. Valent items can, however, be arranged into informational patterns.

That Which We Can See at the Bottom of Our Own Minds from Within is Not the Actual Bottom

We can look down from inside and we can look up from physics. There are two vast regions to be charted. The first is the gap between physics and the lowest things we can see from inside- this span is not of zero size, there is amalgamation and elaboration. The second is all the structure between the lowest things we can internally perceive and the full-blown thoughts and emotions that are the normal coin of our fully conscious life.

Feeling Does Not Require a Separate Feeler

Because we exist of our own nature and at our large scale, people have assumed that- even for "one's own feelings"- there must be an experiencer of "those" feelings. At the bottom, feeling happens [4]. It exists freestanding. It does not happen to something else, at least no further than in the sense of particles mutually interacting in and as the fabric of the universe. This is how the descending homunculi disappear.

Cognition rests on even less. With pure feeling, there is no pattern and the nature of whatever is happening is everything. With information, it matters not what the stuff is that is arranged into patterns, only how it is arranged. This means that with a full reduction the only stuff that is needed is the feeling stuff!

Awareness is Feeling and Not Cognitive

You can feel the presence of something before and without any thinking about it and without even knowing what or where it is. In contrast, a mechanism (including in a brain) can be "aware" of something and respond to it, but it is not doing what we do when we say we are aware of something. The difference is feeling. The same responsive acts performed many times become rituals and fade from awareness.

Attention, Purely, is Cognitive as Opposed to Feeling

Attention is selection of what to focus on. It allows a single difficult-to-evolve unit to process one (or a few) at a time of many inputs, and thereby serve all the inputs. Attention may use feeling as a selection determiner and operate on feelings as the things selected, but the mechanism of switching between inputs is mechanical ("cold") itself. One can, however, be additionally aware of that which attention circuits have switched to, including both one's own thoughts and one's own feelings. In fact, that is the normal case in humans, where attention-switching and awareness are tightly interactive and seem as though a single base function.

The Act of Observation is Not Just Informational

To observe includes to become aware of and become aware of related detail.

Subjectivity is Not Just Informational

The realm of the "subjective" includes both "subjective experience" and "subjectivity."

Subjective experience is not informational, it is the opposite of that: feeling.

Our Subjectivity has been described as an informational relativity. At a very high level and with discipline (which is feeling!) a mind can be trained to consider formal things where this would be so. But that is not at the core of what Subjectivity actually is to a brain and how it works.

A "Subject" is born when a clot of feelings is affected by something outside itself- when, at a higher than fundamental level there is now a distinction between the experiencer and the experienced. When the other something is in the same mind, we have what amounts to one clot of feelings experiencing another clot of feelings. (This can result in bad feelings regenerating further bad feelings mushrooming into a widespread dysphoria.)

In practice most such subjective events for us are further elaborated both at the experiencing site and the experienced site with informational content, giving rise to complex emotions as well as feeling-laden observations and introspections.

Consciousness is an Epiphenomenon and Consciousness is Not an Epiphenomenon

The Hard Problem [3] itself suggests that pain and pleasure 1) serve a purpose that 2) might have been served by some other mechanism in some other universe. They key is to provide valent value serving to steer judgements (at all levels and scales).

The question becomes epi- in what way, epi- to what exactly? The subjective experience Is not an epiphenomenon in the sense that it provides the valent that is used by animal brains to steer cognitive circuits and select and enhance memories. It is an epiphenomenon in that, if there is something that can do the same job in some other universe, that other phenomenon would not have to be feeling.

Circuits can first be steered by this conscious experience and then be written into nonconscious elements once correctly shaped. We see this sort of mechanism when we learn to drive a car and it also happens in evolutionary time periods affecting wiring.

Consciousness is Natural and Physical

Consciousness occurs in our universe; it is natural. And, though crudely, minds are able to interact successfully by way of beings of and operating in the physical world.

Every important and mysterious phenomenon we experience has fallen to a physics description save this one. There is every reason to expect this one will, too, and little reason to doubt that. In the universe in which we find ourselves, we find a describable order in all things and happenings.

To be sure, consciousness is a strange thing, but so were many other now-accepted concepts of science, such as action at a distance and electromagnetic waves and quantum behaviors.

If we bring an electron near another one, the first will scoot away. That would be very strange to encounter for the first time, yet at least that case resembles in some ways our familiar world of macroscale objects.

Since then, we have found quantum behaviors that to our minds crafted for the Newtonian macroscale have been hard to understand, but we have found the many correlations to objects we are familiar with and so have become convinced.

Thus, we have been here before. We are still early in our search for an understanding of consciousness and there is much less to attach to; with a few more puzzle pieces in place, it will become easier.

Feeling is a physical phenomenon that we can expect to be explained exactly as far as the rest. Why does the electron move? We can say "because of a force" but that just defers the query by one step. When we reach the fundamental, the answer is "it just does," which is fine because what that really means is "because that is a part of a system which we know is possible because it exists, and because this particular possible system can support inquirers like ourselves."

We are even more "inside" consciousness than gravitation, which makes the big picture harder to see, but we will have equivalent realizations to that of us being at one star inside a galaxy among many galaxies.

A certain particle interaction emits a neutrino. Why? Within certain constraints, it just does. So, too, with feeling. When we find the link to the movements in our brains, we will be comfortable with the new facts.

We expect that minds are built of very small things and do not originate naïve but whole.

Consciousness is Not Just Information Processing

Consciousness requires both cognition and feeling. While one can't make pain or pleasure from neutral objects, even less can one make them from nothing at all (information). Neither the pattern nor cold pebbles placed to represent it (no matter how complex the arrangement) can make feeling.

The cognitive is 100% equivalent to a simulation; feeling is not (Gamez, Human and Machine Counsciousness, 2018).

The human mind is not equivalent to a simulation for one reason and that is because it *actually* feels. Here we see the difference between an abstraction like mathematics, which can carry information, which is itself simply description of pattern *of* something, and the actual. Consciousness requires the actual.

Complexity Does Not Yield Consciousness

There is a body of argument extant that posits that consciousness appears because brains (or artificial devices) eventually become complex enough. Though as for any machine there will need to be a certain complexity, this is not the source of consciousness.

The position misunderstands systems theory. Systems can indeed exhibit *behaviors* that one would never have predicted by just looking at the constituent parts (though they can be discovered by simulations incorporating models of the parts), but that does not mean a system can exhibit *any phenomenon* if only we arrange its parts with enough complexity.

You can arrange ball bearings all day long and you will never be able to make a living, feeling, sheep out of them. It matters what the *parts* are, what their nature is, what their potential is insofar as what is possible to make from them [8].

If feeling *should* suddenly appear from complexity, then a) we will have discovered not an "explanation" but a new brute fact of the universe, like electric charge, and b) this will very probably most minimally appear at a very small (molecular scale at most) level and with a simple arrangement. In other words, it will be a simple fundamental, not a complexity result!

Like everything else, the consciousness we know arose from something already there of the same basic nature.

Mind is Not Unitary Except Through Fusion of Constituent Parts

The illusion of the mind as a unitary object tends to bias our thinking unhelpfully. Yet, this fusion of experience is also a key clue into how the mind works and especially as to how feeling works in it [4].

It Matters What Stuff the Mind is Made From

Our minds cannot be made or remade of arbitrary materials and mechanisms. There are multiple ways the computation could have been implemented because the material is just placeholders to represent information (although most of these ways ultimately rely on electromagnetism and matter).

But feeling requires that whatever physical mechanism is at play is represented in the brain, by constituents that play that way [4] [6].

The Building of the Machinery for Human Cognition Required and Requires Feeling

While cognition is not about feeling, the mill that performs cognition uses feeling in its operation and in evolutionary bootstrapping.

Feeling has immediate effect in directing suitable flows of information, which may in later stages of evolution be captured in circuits that no longer need to feel. In this case the subjects that are then associated with good or bad feelings are higher-level (or more novel) than were those of earlier times.

Feeling is Everywhere

Animals are organized constructions that harness a few basic behaviors of the universe to enable animal-level actions, promoting competitive genome survival. Minds and what they do are just one more example of this. Minds necessarily had to be formed from what was, in disorganized form, already present.

There is No Conflict Between Free Will and Mechanism

At times hand-wringing over Free Will has entered the quest to understand consciousness. One does what one wants to do, all things considered, based on present and anticipated pains and pleasures. This can be stated either as a definition or an observation and is true in both cases. In fact, how your brain reacts to all the various pains and pleasures in making its decisions is what your personality is [9]. Your Will is what you want. Like all other bodily functions, it is a fact of nature, and we are literally wired that way. Will is the drive away from pain and toward pleasure.

"Free" is a relativistic term. One may be free from other people's control. As a creature of the universe you are not free of its control. It also enables and implements you.

Procedural: Incrementalism Plus Empiricism Will Not Discover the Answer First

Next, we move to cautions about procedure rather than natural truth.

The correct answer for the puzzle of consciousness will come originally from advanced imagination and suspicion, to be later confirmed by experimentalists. In particular, the notion that if one cannot now detect something with the instruments we now know how to make then it probably doesn't exist, is woefully impotent today on the topic of consciousness. "Laws" (including conservations) have been broken on the way to the Standard Model; without being willing to consider these possibilities we would not have reached our successful Model. Consciousness is stranger still than earlier challenges and leaps are necessary; compiling a compelling body of evidence experimentally to demand one answer would take a very long time, measured, as one wag put it, in the unit of dead scientists.

Consciousness takes us into BSM (Beyond the Standard Model) physics- we know this from the start.

Procedural: Soft Laws are Valuable

It is not certain that the laws of consciousness will much resemble the laws of the other physical phenomena; there is no logical proof of this and there is no empirical proof today, either. However, two things push us to that plausibility, and both are based on general experience only. The first is Occam's Razor; we have found that the correct answers tend strongly to be among the simplest possible answers. This is one manifestation of a second general and true rule of our universe that appears ubiquitously, which is that that which has been found to be the case many times will probably appear repeatedly again.

Procedural: Excess Aversion to Addition is Counterproductive

It is proper (wise) to minimize in a new theory the bringing into being of new objects and concepts with little empirical evidence to back them up (an ontological aversion).

However, it is a mistake to meticulously avoid this *when all other avenues are yet more contrived and arbitrary and implausible*. The bar should be high, but not infinitely so.

A true Theory of All (TOA) will be Standard Model + Gravity + Consciousness. There must be enough conceptual material to cover that span.

As with consciousness we have other mysteries such as dark energy and dark matter, and ordinary matter and gravitation. As the later pieces of the physics puzzle are filled in, multiple mysteries may fall at the same stroke. Fundamental consciousness may be related to or even identical to one or more of these. Consciousness may turn out to be not only a universal stuff but *the* universal stuff, an inversion of our prior thought paths.

With no more actually required than the observation that all other forces have messenger particles scientists have been justified speculating on a graviton. Similarly, an inflaton to explain cosmic inflation. Einstein's contributions were often arguments of logic made from a small number of basic empirical facts; his mathematics of gravitation were one possibility only later upheld empirically.

Metatheories- that is, bounded collections of theories- are a sensible vehicle on the journey, and these may even be neutral on questions re new objects. A logical course, for example, is to point to and name a new particle, with the explicit proviso that it may actually be new, or a familiar particle used by Nature in a previously unknown way (like electromagnetics being married to electrostatics).

It is not only possible, but likely, that the physical description of consciousness will involve both familiar and strange structure.

Procedural: Sense of Plausibility is Important

While there is little to go on, speculation is necessary and skill in recognizing the plausible (especially where change is required) increases in value.

Consciousness theories exist featuring no discussion on *why it* may be a likely true or nearly true theory or metatheory. Many things could be true- which are? This is a hard type of evidence to express (and it is evidence)- it relies on intuition and judgement not necessarily felt by the shaped neural nets of others, but it is precisely these kinds of leaps that first speculate the correct paradigm shifts (this we call "genius"). Timidity does not serve at the frontier.

Epilogue: A Path Home

How do we solve the mystery of consciousness? High enough analysis and in the right way, "lower than philosophy, higher than our standard physics."

We have physics laid down now that mostly helps us but also tricks us by asserting patterns that may not entirely hold for something as strange as consciousness. One excellent clue from physics that probably does hold, is that everything fundamental about consciousness probably does not require large expanses of space or time.

The first step is to realize that feeling is an action, and apparently a fundamental action. Given that it is a fundamental action that is not the same as the others, as is the way of our physics we posit a field for it as well. It follows that there is also a charge and at least one associated force particle [10].

We frame this as a metatheory, admitting that we don't yet know the specifics of this field, nor its relation to the existing ones. We acknowledge that "new" particles could be existing particles about which we have incomplete knowledge. Nonetheless the first logical position is to frame the theory in terms of fundamental consciousness without commitment on these finer points either way. This is not falsity, it is generality.

Until we are able to do empirical experiments that are shown to have access to consciousness (FC), we must rely on informed speculation with an emphasis on plausibility, and thought experiment, and avoidance of the easy pitfalls discussed.

We must recognize that it is far from certain that the laws of consciousness respect, for example, special relativity, or operate mostly in our familiar dimensions of space and time.

We should maintain the obvious suspicion that consciousness may have something to do with- perhaps a great deal to do with- collapse of the wave function or gravitation or dark energy or matter as we know it.

We must entertain theories that have less empirical support than we have been comfortable with. We must recognize that while most imaginings are false, a few are true. We must mine our sense of the plausible because our understanding of the truly required is unable to serve us sufficiently here.

We must create metatheory and catalog the metatheories and see how precisely we can describe and bound them, despite the risk that entails.

We must appreciate complexity layering from fields and particles up to animal minds and come to see how this can happen.

The degree of explanation of the forces we know is just a descriptive catalog of which particle combinations interact and how strongly and quickly and in which directions. In the absence of any strong evidence to the contrary we should assume the same posture with consciousness.

What we know so far suggests that consciousness is a) pervasive, b) very fundamental

and c) that we don't just have it, we are made out of it- how much so is an intriguing question.

How to prove is an important but separate problem from figuring out a plausible approximate explanatory theory or class of theories (metatheory). While we are unfortunately missing means to test, we should not idle ourselves in waiting, and likely the very hard brain work of the present era will suggest the experiments. Many of the really important items we have found were not discovered until after we developed better provisional ideas of what to look for.

There are interesting precedents as humankind realizes and proves (in that order) how ever more deeply we are made of that which surrounds us:

We are animals, too.

We are made of cells like all other life here.

We are made mostly of the water that everywhere surrounds us, and emerged from it.

Our Sun is one of the stars.

We and the Sun are made of older stars.

Now consciousness we understand is an energy of the universe and not somehow just found in animal heads. Now we find ourselves in a universe laden with dark matter and infused with dark energy as our minds ponder what "other" things these are, and what feelings are, and how exactly probabilities and desires become actualities. It will be strange indeed if we do not find yet deeper kinship in nature.

Conflict of Interest Statement

There is no conflict of interest for the work of any article written by me through the year 2020.

References

[1] R. DesCartes, Principia philosophiae, Amsterdam:: apud Ludovicum Elzevirium, 1644.

[2] G. Edelman, Bright Air, Brilliant Fire, New York: Basic Books, 1993.

[3] D. Chalmers, "Facing up to the problem of consciousness," *Journal of Consciousness Studies,* pp. 2: 200-19, 1995.

[4] K. Sipfle, The Nature of Consciousness, New York: Barnes and Noble Press, 2018.

[5] R. Penrose and S. Hameroff, "Consciousness in the Universe," in *Consciousness and the Universe*, Cambridge, MA, Cosmology Science Publishers, 2017, pp. 8-47.

[6] R. Pozonanski and E. Brandas, "Panexperiential materialism: A physical exploration of qualitativeness in the brain," in *Quantum Boundaries of Life*, Cambridge, MA, Academic Press, 2020, pp. 301-367.

[7] D. Gamez, Human and Machine Counsciousness, Cambridge, UK: Open Book Publishers, 2018.

[8] N. Chomsky, What Kind of Creatures Are We?, New York: Columbia University Press, 2016.

[9] E. T. Rolls, Cerebral Cortex: Principles of Operation, New York: Oxford University Press, 2016.

[10] G. Kane, The Particle Garden, New York: Basic Books, 1995.

The Solvability of The Hard Problem of Consciousness and its Relation to Fundamental Consciousness and Human Consciousness

Karl Sipfle

ksipfle@umich.edu

April 2, 2019

"feeling is first who pays any attention to the syntax of things"

-E.E. Cummings

This presentation is very terse; amplifications can be found in (Sipfle, The Nature of Fundamental Consciousness, 2018) (Sipfle, Distinct Kinds of Consciousness, 2019) (Sipfle, Support for the Sentonic Theory of Fundamental Consciousness (STFC), 2019) (Sipfle, The Nature of Consciousness, 2018).

The Hard Problem is two sub-problems:

1. How and why there is any subjective experience at all? What is the source of this?
2. How and why do we experience many different qualities that appear to be basic yet exist only in minds, such as redness and pain, without any discerned link to aspects of the physical universe?

Full solution of this Problem can be had in steps. The first is to demonstrate any one way in which subjective experience could be connected to our physical universe (which is the scope of this article). This way must be plausible insofar as it obviously shows itself to be a member of a set of similar theories, such that the set is itself plausible for a universe such as ours. It is not especially important, even if possible, to identify every member or characteristic of the set. This proves solvability, and in the process offers candidates for the correct actual solution in our world. Presenting a possible solution likely comes before and is helpful toward *the* solution, which may have similarities.

The second step is to devise means to experimentally test the original theory and its variations and do so.

Assumptions

This paper makes a few assumptions, which are out of its scope to justify (see References).

We presume physicalism and reject information- and computation-based theories for the source of conscious experience.

There is a physical consciousness force, accounting for *fundamental consciousness* (defined shortly).

It is probable that consciousness functions by the same rules (including mathematics) as all other phenomena that also exist in Nature.

Animal-level consciousness is probably the result of activity of networks of physical neurons.

Levels of Consciousness

We divide the vast scope of what is called "consciousness" (adapted from (Sipfle, The Nature of Consciousness, 2018)):

Level 1 Consciousness a.k.a. Fundamental Consciousness a.k.a. Base Consciousness

This consists of the smallest elements that represent experience. (This is where free-standing experience occurs, without an observer.)

Level 2 Consciousness a.k.a. Sentience

Consists of everything that must be added to level 1 to make the sentience of the simplest sentient organism.

(This is where "subjective experience" occurs, and it introduces the concept of an observer.)

Level 3 Consciousness a.k.a. Mind

Contains everything above level 2 that enables the behaviors and experiences up to the most complex of nonverbal organisms (by adding cognitive componentry).

Level 4 Consciousness a.k.a. Human Consciousness

That which has been attained by humans (by adding language) (Edelman, 1993) (Rolls, 2016).

Level 1 Consciousness plus Level 2 Consciousness are the foundation beneath the entirety of "consciousness" defined as "a bubble of experience" if that bubble "consists of colours, sounds, smells, tastes, etc." (Gamez, 2018). Level 1 Consciousness has its own elemental bubbles of experience, which we may call "psybits," but the experience is more primitive and not in all occurrences integrated into any kind of mind or substantial piece thereof.

We will see that Level 1 is the root of actual experience, rather like Layer 1 of a computer network, the Physical Layer, moving around real electrons. In both cases, all the rest sits on top and is connected assemblies and software built on Level 1. If we simulate Level 1 and Level 2 and then put all the same algorithms atop them as we find in natural brains, we can get all the same behaviors, but that "mind" will not *actually* feel.

Level 1 Consciousness is the fundamental consciousness that is the physics manifestation of what is needed to build, from it, consciousness as we know it. The nature of this low level and the connection from it upward is the Hard Problem.

IV	**Verbal Mind.** Abstracting Cognitive	More wiring and Algorithms
III	**Mind**. Cognitive	Wiring and Algorithms
II	**Sentience.** Qualia Affective + Cog	"Box 2"
I	**Fundamental Consciousness**. Particles, Fields Affective	"Box 1"

Figure 3: Levels of Consciousness

High-Level Approach to a Solution of the Hard Problem

We can approach the Problem by noting that while we don't know for sure the details within them, we do know the perimeters of two boxes that together comprise the whole Problem, and a coarse architecture for Sentience (Level 1 plus Level 2 Consciousness).

We can draw the first box because, while we don't know exactly how neurons generate base consciousness activity, we know that they probably do (and can leave the absolute proof of that to be done by the experimentalists).

We can draw the second box, because while we do not know exactly how we get from base consciousness events to the so-called "raw feels" at the level of our minds, we know this is probably happening (and probably as some kind of elaboration using Level 1 Consciousness).

The two-box approach lays down a central hypothesis itself, and the internals of each box can then be refined into further hypotheses.

Fundamental Consciousness (Box 1)

Level 1 Consciousness is Fundamental Consciousness (a.k.a. base consciousness, where means "base" from the point of view of the physical universe). Our Standard Model of Physics plus Gravitation theory do not once mention or show any underlying support for consciousness, which we know occurs, so they are incomplete. An additional aspect of the universe awaits discovery to provide the additional medium of which to make conscious minds; it may come as a deeper understanding of things we have already seen.

A single speck of the new phenomenon of indivisible physical consciousness we call the "psybit."

A proposed solution for Box 1 is the STFC (Sentonic Theory of Fundamental Consciousness) (Sipfle, The Nature of

Consciousness, 2018), which anticipates a new physical field. We will see below that analysis of the second box suggests that the new physical aspect will come in at least two variations or states (probably in addition to position).

For many reasons (Sipfle, The Nature of Fundamental Consciousness, 2018), we believe Level 1 Consciousness

Is free-standing, with no observer;

Operates at the level of physics;

Is not generally directly accessible to our minds, rather, most of the lowest level things we can experience are already elaborations built out of Level 1 consciousness.

There are two or three exceptions to the last point above. Directly experienceable in pure form by our minds are pain, pleasure, and perhaps, unqualified impact.

Qualia (Box 2)

Qualia are the hard part of the Hard Problem. Qualia are at the bottom edge of human mind (which is a whole represented by intense interconnections both mentally and physically. The bottom of the mind is well above the physical level making it possible, that is, base consciousness.

In analyzing this box, let us first consider pain and pleasure. These are interesting as subjective experiences that are included as basic experiences to every human mind, but that do not demand a substantial range of possibles, as qualia (such as colors), in general, do. Suffering and pleasure are distillable in pure form from the human mind, and there is only one essential member of each of those two experience classes (or just two of one class if you prefer), not a spectrum or family.

Next, we consider, in contrast, *red*. Red is *not* a single raw feel. Red, and anything else we calmly notice, has *impact*, a feel shared by many other "raw" feels. Furthermore, it is a visual (possibly envisioned) experience, and additionally to those two facts, it is also red.

Impact is like pain and pleasure without the sign. Impact is neutral oomph. (Impact may be a composite of negative and positive.) Minimum awareness is naked impact.

This quale- red- may be the least at once that the mind can do with it, and in that way this supposedly elemental thing may be atomic to the mind, but it is already clearly made of multiple things. Commonly mentioned qualia are not the rawest of raw feels; they are more complex.

Such a minimal, actual experience for us is likely to be an *assemblage* of some kind of that which applies at the elementary physical level. Next, we need to show a believable way in which this could be so.

How do we feel red and blue? A little piece of your brain will light, consciously, when blue light comes in. However, this piece in isolation would not influence the connected whole of consciousness that we call the mind; it would be an outlier. This piece of brain is connected closely with a piece of brain that activates for any such visual color input (these connections are either direct or effectively so). Now we have a mob operating in concert, the experiencer of blue, and the experiencer of visual color upon which other color experiencers converge. This explains both the specific and the more general experience which are happening together. Yet higher convergence is to vision as a whole (fusion). These elements of experience are integrated into a whole, yet with introspection we can detect that both a common visual awareness and a specific redness are occurring.

We must next address an even more difficult issue. What then makes the activity of the red brain patch different from the activity of the blue brain patch? One can say they are separate and so can hold separate information but that is only half the answer. Furthermore, there is nothing at bottom to us that feels special about red or blue, yet they *feel different*.

The two patches of brain tissue will have different microdetails physically, which could cause differing blobs of fundamental consciousness actions (psybits) in space (with these psybits in contact with one another by virtue of their field). It is not important on this point what the exact arrangement within these blobs is, just that they differ. The degree of difference is yet more pronounced if fundamental consciousness comes in positive and negative flavors, because there would then be not only an amplitude shape in space but also a pattern of +/- psybits within the blob.

What is happening is that the detailed tapestry of individual spots (psybits) has different detailed composition, so the feeling of the little mob is different. This effectively gives differentiating identities to each particular feeling; the experience of each is a little different (these then become associated in groups at higher levels of the mind, into richer experiences).

In a signed theory, a typical glob will have *impact* but not pain or pleasure because the plus and minus psybits cancel each other out *overall,* but each *pattern made* of the individual psybits exists and is different.

Discussion

We may say redness is a quale, but it is only a quale to our minds because of the other present components, the common visual impact. The red brain section we might be able to stimulate with a probe but that would not result in a quale in the mind unless the other tissue were still connected.

Red is not a freestanding feeling (pain is), red is something you are aware *of*. All the way up at the level of red, the concept of an observer plus an observee has emerged, which does not exist at the physics level. At the level of red, the possibility of an experiencer of something else (the beginnings of mind) has emerged. This is not Level 1

Consciousness. It is the top of Level 2 Consciousness and the bottom of Level 3 Consciousness.

Finally, note that pain and pleasure take a shorter route to experience in our minds; there is little or no elaboration. Precisely because they are not elaborations beholden to an elaborating process, they provide a powerful, very basic, and general value mechanism that cuts right through the complexities needed for the purely cognitive (informational/computational) functioning of the mind.

Thus, we can (and should) imagine a way to get all the way from particle physics to qualia of red and blue, that is, from the "atoms" of the universe to the "atoms" of mind, without a trivially short and outlandish connection drawn. The forgoing describes not just a solution of the Hard Problem, but a model for other possible solutions to spin off from.

A glob of psybits does not have an experiencer observing an experiencee. This is replaced by the physics influence of one charge upon another. Any two such particles are together the chicken and the egg. The oddly-patterned glob "speaks" as a choir and "hears" as a choir with the rest of its physics environment. It is a molecule of experience.

Why have we not detected base consciousness from within our minds? This is not very surprising given that this consciousness, the tortoise upon which all of thought rests, is even invisible to us in our laboratories, from the outside.

Software (and the higher hardware) cannot be consciousness. However, everything above Level 1 is implemented by software and higher hardware. Everything above Level 1 and Level 2 is not really consciousness at all. Whether considered alone or running on other hardware or simulation of hardware, it is just a simulation. "Everything about a man that is a man" (-A. MacLeish) *is in the base hardware* (the very place some say we should not trouble ourselves about), and everything that is the most mysterious and important lives there. The rest is just a fancy algorithm almost certain to fall to computer science soon.

Gamez (Gamez, 2018) says "It makes little sense to say that colourful smelly noisy bubbles of experience are identical to something that is invisible, silent and without smell. This discards the properties of bubbles of experience and ignores the reality of our day-to-day world." This statement turns on the word "identical," and so is of limited coverage, even if taken a little more liberally. It deflects our gaze from where we need to look. The answer to the concern appears to be that the smells and colors that we find so special against our backgrounds are simply distinguishable little blobs of fundamental consciousness that have become associated in our minds with a great many things- a great many other blobs. For example, you say that you can experience red. What does red feel like? The answer is it feels like what it feels like, there is nothing more you can say, even in principle. It is a

distinguishable experience that can be (and is) linked to others.
We do not expect a gear to tell us the time, but we do expect gears in assembly to provide a clock that can tell us the time. Exactly when we see the assembly, it all makes sense.

The question becomes, “How do we get from sterile little balls of very few kinds to different qualia?” The two available approaches to an answer, in principle, are:
1. Each different kind of quale is underlain by a different fundamental in the universe. The universe knows how to feel and distinguish red and a hundred other things.
2. The supposed fundamentality of any quale of human minds is one of many elaborations made out of smaller bits (and with many other types of qualia being possible in principle but not present in us). Obviously, this is more likely, whether or not we understand how.
In other words, the usual qualia are *not* base feels; they are more complicated constructions.
Because of all this we will not be able to reduce red qualia to a single characteristic particle movement. But we should be able to identify pain with a single characteristic particle movement.

Let us invoke the bat (Nagle, 1974) and remember that the question at this level is not what those complexes of globs feel like “to you.” *They* feel, those little bats, and with enough of them arrayed just so, they *are* you.

It is possible that besides providing value, and an additional means of communication, fundamental consciousness provides persistence- memory. This might be especially important to our experience of the physical property of time.

Levels 3-4 comprise entirely the cognitive portion of the mind and (though fascinating) are completely insufficient to explain why one could have a what-it-is-like (Nagle, 1974) at all, which is *the* thing that is really interesting and different and special about consciousness (and us). Efforts to avoid (actually, postpone) the physical details so as to avoid wild goose chases, poignantly avoid (postpone) precisely what is needed to understand the thing! An outside-in mission plan and noble rigor may work, however, to bring us down to the answer from 100 km out, by striving to stay on course.

On Approaches, Methods, and Scales

In explaining all of our real conscious experience, the lowest portions are a physicalist situation- Feeling. With this platform in place, the remaining, higher portions are simply a computational situation- the Cognitive.

Two mis-directions often made both come from minimizing one or the other.

Technically the whole potato can be deemed physicalist, because that element is required, and because any arrangement and motions no matter how elaborate, can be claimed to be more description of what the physical world is doing. But in truth the bulk of the mind is algorithm and storage atop the basics of creating mental experience. Once this latter detail, as some have called it, is functioning, the rest is all computer science and experimental psychology.

The other error, by being biased the other way toward the atoms of the *mind*, is in presuming that these atoms represent the tiny things at work. Indeed, the fact that qualia don't seem much like particles has been used to deprecate particle-centric pursuits (Gamez, 2018), but qualia and particles look so troubling and confusing in their difference only before realizing your quale is "a cabbage and not a cell."

Gamez defines "consciousness" as "a bubble of experience" that "consists of colours, sounds, smells, tastes, etc." and uses the term "physical world" in exclusion of this consciousness. He divides the Hard Problem into three parts:

1. A Pseudo-problem: It is impossible to imagine the relationship between consciousness and the invisible physical world.

2. A Difficult Problem: We find it difficult to imagine the connection between conscious experiences of brain activity and other conscious experiences.

3. There are brute regularities between consciousness and the physical world that cannot be broken down or further explained.

On #1 we disagree, as we have just done so.

#2 is unimportant given the definition of consciousness in use here, "bubble-consciousness." But expanding "consciousness" to include fundamental consciousness, this connection is not only imaginable but profoundly important. It potentially includes a recognition that conscious experience happens as an elemental, freestanding event spread as motes (psybits) across the universe and constituting the second type of "stardust" of which we are composed.

At some point we will get to the irreducible brute facts, but this point should come at the level of fundamental physics.

Gamez proceeds to build a framework for the study of consciousness that essentially draws the known perimeter of the battlefield and demands evidence, to drill down to the essential physical level from above. But the answer as to where consciousness comes from, in particular the defining subjective experience of it, is very likely to be extremely dependent on and similar to the nature of Nature as we know it, and good speculation can tease out new Pilot Projects.

Bubble-consciousness requires Level 2 to make qualia given fundamental consciousness, and the first bit of Level 3 to form a single-bottle mind in which many of them can appear. Everything above is Level 3 and Level 4 and is software and higher-level hardware. That which makes it all real is what the "bubble" is implemented upon and out of- the physical- which is excluded from the "bubble."

Gamez supports physicalism, but knowingly terms the "physical world" to be the world that is not "consciousness," looking for the link between the two, with the new physical understandings represented at first only in those links between the two worlds. That construction, a different parcellation of same concepts, can serve well.

It is unconvincing, says Gamez, that observing neural patterns explains qualia. While it is indeed unconvincing that the existence of a field all by itself can immediately explain a feeling of red- there is not enough there there- to see the explanatory picture does not require a "wave of the hand," (Gamez, 2018) but a construction of the mind. It is also very believable that something at so low a level as a particle or field *can* be experience of a *really* basic feel like pain or impact.

We know that feeling is. At some point, stripped and reduced, it still is, and at most can be further explained by its association with something else also primitive, to then be elaborated, as with everything else in our universe, into things we, at our level, more easily understand.

While one cannot reduce the traditional "fundamental" experiences of our minds to a trivial mapping to features of the physical world, one also cannot do without these same features and still have the full theory. Such a theory requires the particles that play at the physical level as well as the large hubbub in which they are participating. Smells cannot be reduced to single particles; what has been oft-missed or glossed over or, paradoxically, seen as a problem with quantum physics-oriented approaches, is that smells are not elemental as far as the universe concerned. Particles and fields are not eliminated as the real actors and they probably are the real actors. In fact, without the particles to assemble into their globs, it becomes a very abstract undertaking indeed to predict the correct globs and their composition.

As Gamez points out, the outside-in approach risks getting stuck at another intensity of mystery once certain things have been characterized but have no path apparent to break down further. Then we will just have another Hard Problem. The value of doing so anyway is that we can start collecting more hard data on the journey. Where we are light on explanation, we will also have high confidence in what we do know. While important, this is not by itself the answer. Someone is going to have (or has had) a blinding flash, having looked at the problem from many views, that provides missing "details."

Finding a path all the way through the hard problem was enabled in our case by starting from the field and particles physics we already know and asking how we will get to where we need to get. It will be surprising if *nothing* in the theory presented here matches the true theory that is proven in the lab.

The operation of real material in our world demands both atoms and all the complexities of the separate-yet-joined-into-wholes way in which these atoms participate, and this looks to be almost exactly what is happening with our most basic experiences, as we should expect. A cell *is* made of atoms, even if upon hearing about atoms you still don't see how that could explain life. That just means there are more steps of explanation and of structure than are given with a simplistic hand-wave. What is required is not just to avoid oversimplified explanations but also oversimplified presumptions of what we are starting with.

The true brute facts occur at the bottom and amount to nothing other than the rules from which the universe is made out of its energy. There has to be *something* standing in that place. These will include the Physical Correlates of Consciousness that underly and explain the Neural Correlates of Consciousness.

The pursuit of an understanding of the operation of minds such as ours can be rather neatly divided into two teams of investigators- one studying how basic feelings come to be and one studying how these fundamental experiences can then be used to build mental machines. It is likely that major progress can be made in each before there is a complete sewing-together of the two theories (as a similar example, the Standard Model of the small scale and General Relativity of the large scale come unavoidably to mind).

Summary and Conclusions

A fundamental consciousness operating at the level of quantum physics underlies all conscious experience. At this level there are only particles affecting each other, consciousness is a freestanding physical event, and there are no observers present.

Typical qualia occur at a higher level and are elaborations incorporating "psybits" of fundamental consciousness. That qualia are themselves composites is shown by introspection and by the realization that, otherwise, the universe would need a hundred different essential flavors of consciousness, an explanation which is both suspiciously bloated and very unlike the rest of the known universe. Additionally, varietal compositions (based on forces) is how all the rest of the known universe has organized into higher structures.

Since psybits affect each other at the physics level, so do the "molecules" that make up qualia.

Components of different qualia have distinguishable feels because the molecules of psybits have different exact makeup and internal arrangement, made inevitable by variations in the physical brain.

If we gradually anesthetize a brain, human consciousness eventually disappears, at a point before almost all neural firing has stopped. Likely, the longer and thinner links between locations of high self-supporting connections break first, leaving islands of independently conscious chunks of brain that no longer result in a unitary mind. As activity continues to fall, these islands of activity with their internal tighter loops will also collapse. At the most extreme nonzero case, sputters of fundamental consciousness will be evoked, but these will be both too weak and too simple to amount to any anything mental. They will be simply more noise in the universe.

A simple experience to us involves many neurons supporting each other's firing, each (conscious) one of which probably cause many individual physics-level events. A simple thing to the mind is large on the particle scale. Even a weak feeling to a person is rather rich in absolute terms.

What we call our consciousness is that which is accessible in the unified mind. Outlier bits of consciousness will exist in other parts that are not connected in, in our own brains, other peoples' brains, and probably in the whole universe. The boundaries of the unified mind are probably dynamic as things enter and leave our "consciousness."

With all of the feeling in place, all of the rest of the mind is built by Nature atop this platform- minds have evolved, which are mostly information processing, and which would not actually feel were it not for the nature of the lower levels that provide the medium.

Researchers who see consciousness as fundamental in the universe without recognizing that it is assemblies that make qualia in minds, and psycho-philosophers who think their red is a natural primitive, are talking past each other in trying to make the connection between their worlds. Physics is not high enough and psychology is not low enough for that. Even introspectively, we can see that red is not an irreducible sensation, and neither is redness a moving particle.

We have seen that the large question as to what a whole human mind is made of can be well-split into two pieces, which are basically the Hard Problem and a lot of information computation and storage. Therefore, the solution may be sought by two different specialties and teams of humanity. Realizing how naturally separate they are serves to make each search easier. As to applications, understanding of the feeling half will have little importance for the creation of effective artificial (and stoic) “minds,” but be of central importance to extending our own conscious lives forward.

The key link in understanding the whole of consciousness is the explanation from physics to the much higher level of qualia, at the bottom edge of the mind. Key clues might be found in pain and pleasure, which appear to have a much simpler and more direct link “upward” than the typical qualia.

The described model shows both a conceptual solution to the Hard Problem of Consciousness and a candidate, with likely refinement, for the truth, to be proven by experiment.

References

Edelman, G. (1993). *Bright Air, Brilliant Fire.* New York: Basic Books.

Gamez, D. (2018). *Human and Machine Consciousness.* Cambridge: Open Book Publishers.

Nagle, T. (1974). What is it like to be a bat? *Philosophical Review*, 435-450.

Rolls, E. (2016). *Cerebral Cortex: Principles of Operation.* Oxford: Oxford University Press.

Sipfle, K. (2018). *The Nature of Consciousness.* New York: Barnes&Noble Press.

Sipfle, K. (2018). The Nature of Fundamental Consciousness. *RG.*

Sipfle, K. (2019). Distinct Kinds of Consciousness. *RG.*

Sipfle, K. (2019). Support for the Sentonic Theory of Fundamental Consciousness (STFC). *RG.*

The Emergence of Qualia and Concepts from Physical Fundamental Consciousness, the Gross Architecture of the Mind, and Implications for Further Study

Karl Sipfle, Independently

At NASA Johnson Space Center

DRAFT REVISION: September 14, 2019

The focus of this paper is how physical fundamental consciousness events can be elaborated into mind-fundamental consciousness events (qualia), providing a platform for mind as we know it through purely cognitive improvements. Building from the bottom (the realm of physics), this paper proceeds up to the top surface of Consciousness Architecture Layer 2, which provides the full substrate of mind upon which the full mind can be and is entirely made by adding only computational elements, namely wiring and software.

Definitions

Consciousness: This term has many meanings in use. We divide into four Levels (Sipfle, The Solvability of The Hard Problem of Consciousness and its Relation to Fundamental Consciousness and Human Consciousness, 2019):

Level 1 Consciousness a.k.a. Fundamental Consciousness a.k.a. Base Consciousness

This consists of the smallest elements that represent experience. (This is where free-standing experience occurs, without an observer.) Because of the variation in use of terms, Fundamental Consciousness has sometimes been termed Pre-consciousness.

Level 2 Consciousness a.k.a. Sentience

Consists of everything that must be added to level 1 to make the sentience of the simplest sentient organism. (This is where "subjective experience" occurs, and it introduces the concept of an observer.)

Level 3 Consciousness a.k.a. Mind

Contains everything above level 2 that enables the behaviors and experiences up to the most complex of nonverbal organisms (by adding cognitive componentry).

Level 4 Consciousness a.k.a. Human Consciousness

That which has been attained by humans (by adding language).

Cognition (adj: Cognitive): Thought, as opposed to feelings.

Perception: Informational understanding (not emotional).

Qualia: "Raw feels" that people have claimed to exist through introspection from within a mind.

Valence: Goodness/positivity or badness/negativity.

Emergence: A consequent complex whole arising from the interactions within a system, that takes on its own characteristics not describable in terms of the individual elements of the system.

Blind Alleys

Here we clear the deck by dispensing with some persistent mistakes.

Very Large-Scale Quantum Effects

Theories of large-scale entanglement for consciousness lack both physical and, especially, persuasive argumentative evidence.

The Myth of Complexity as Source of Consciousness

Some have tried to ponder emergence philosophically and turn it into a question. Perhaps using a different word would help those discussions, but here we are dealing with systems, and emergence is nothing more or less than what happens on a larger scale by causation from the properties and behaviors of the components. What emergence is not, is a ghost that can make something entirely new and different out of nothing or nothing that is in some way related.

Emergence is a quality of systems made out of parts, whereby the interactions in combination can produce behaviors from groups of the parts that become characteristic of the group and are not describable from just the level of the part. To take a trivial example, let us suppose that people in a stadium who see everyone to their left rise vertically to their feet for a few seconds then tend to rise vertically to their feet for a few seconds. What will result from this is a large wave horizontally circling the stadium. *This is a consequence of the individual behaviors.* This is system-emergent behavior. It becomes possible because of the complexity present- many people are present, interacting- but is not a result simply of complexity- the people and their behavior is absolutely necessary (and sufficient) to create the higher phenomenon.

A brain and its mind will have much more complexity than that between them and the quarks of the universe, but the principle is the same. In reality, to emerge is to emerge *from* something, without magic. Often one cannot predict before the fact what the behavior will be, but that does not change the fact that it was caused by what made up the system. Consciousness at the human level does emerge from simpler things, but there is an unbroken causal chain upward; there is not a spot in the middle where something just happens purely because things have gotten complicated and for no other reason.

One is never able to make a pig out of ball bearings or bottles or paper towels (or pure information). The constituents have to bear some relationship to the whole in order for the whole to emerge.

Consciousness is not Just Information and/or Computation

A mind does two things: think and feel. Fancy software can (at least primitively) think. It can simulate behaviors related to feeling, too. But what it cannot do is feel-actually feel. That is because it is only manipulating information and all known computers represent and manipulate that information in a totally different fashion than do biological organisms.

"Information" is an abstraction; it means the patterns *of something*. It turns out for consciousness it matters a great deal what the something is.

An object that can do nothing but feel pain we acknowledge to have a glimmer of consciousness. An object that can do nothing but process information we do not take to be conscious.

Qualia, and true raw feels, which are different, are not Information and are not Complexity. The maximally wrong approach sees consciousness as an emergent phenomenon from information processing.

The Gross Architecture of Consciousness

The first thing we should note as to the emergence of qualia from physical fundamental consciousness is whether there is any difference between the two. The answer is Yes, and this is terribly important to understanding of mind. In fact, there is also a large difference in nature between the different qualia.

The identification and classification of Levels of Consciousness corresponds relatively well, it turns out, with layers of architecture of real mind. These are shown in Figure 4: Architectural Layers of Consciousness.

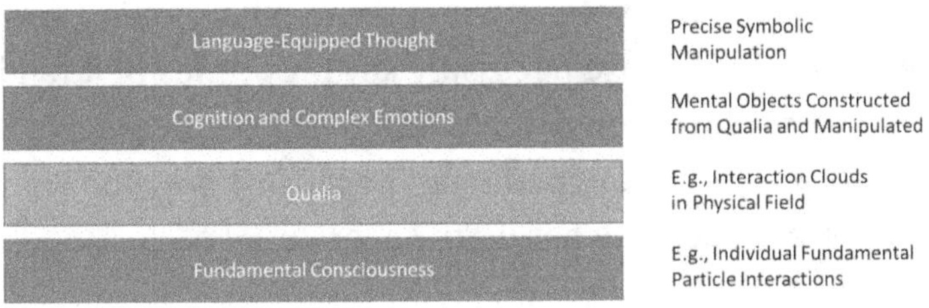

Figure 4: Architectural Layers of Consciousness

These architectural layers of mind include, from the top down:

Language, integrated into thought.

Cognition, integrated with feeling.

Higher manipulations

Manipulation of base objects

Qualia, simple and complex.

Fundamental physics, from whence come base feeling and objects that can be used to record information.

Consciousness Architecture Layer 1

All phenomena we humans have identified have been seen to emerge from things that appear at the fundamental particle level through, maximally, the molecular scale.

Though we are not yet certain of the mechanism in Layer 1, we can characterize necessary key properties of it (and thus also speculate on actual mechanisms). This allows us to describe, without impediment because it is supposed to be more abstract, the nature of Layer 2.

Problems that must be faced in getting up to qualia (Layer 2) include

6. Qualia are not all of the same character
7. Qualia are multitudinous
8. A feeling is fused into a whole, and more intense feeling is a stronger whole

9. At the bottom there are no Observers (in the mind sense), only observers (in the physics interactions sense).
10. There must be a way- a path- for our minds to have evolved

We now discuss Layer 1 from the standpoint of what it must be like in order to support an explainable Layer 2.

Layer 1's province is the true indivisible, smallest feeling that can occur in the universe. Pursuing that leads us to some important (and non-traditional) conclusions.

First of all is the problem: for a feeling (commonly called subjective experience), who is feeling it? The process of drilling down to physics reveals that the question itself is essentially a meaningless assumption when we get to physical fundamentals rather than psychological concepts. When two electrons collide, which one is the observer?

> Principle: Feeling in its simplest form does not happen to an observer, it simply happens, as a freestanding event. The whole notion of a subject to experience something is a much higher development (and a true emergent phenomenon).

The Observer as we know it comes late in the upward process. The Observer amounts to one clot of feeling and computation taking as input another clot of computation (and often, feeling). The Observer is a swelling in the unified corpus of mind, and is not even 100% separated from the (internally) observed or 100% the same thing from one moment to the next.

At the bottom there is no observer, at least not one that is somehow fundamentally different from the observed. Electrostatic interactions are not Observations, in a mental sense, they just happen. Fundamental feeling is the same way.

Secondly, all of our physics describes very well everything in the so-called "physical" world, from the tiny to the immense and everything in between, while not at all describing such things as pain or pleasure. This tells us that

> Principle: Something *basic* is missing from, and must be added to, physics, which is our description of reality.

Our feelings are real. Any Theory of Everything that really is that must include feeling or a way to make it. That the problem may be hard to solve does not make it invalid or meaningless as a problem. (Ironically, the only reason people study and develop physics in the first place is that they *want* to.)

The missing simplest physical feeling event we call Fundamental Consciousness. Individual occurrences we call psybits; each is a "microquale" ((Poznanski, 2019) refers to equivalent "preconscious microfeels").

Notice that, apart from the fundamental physics nature of the feeling interaction, there is not necessarily anything especially quantum about this (although it may involve interactions that include "additional" dimensions). On the other hand, because the fundamental (quantum) level of Nature extends up to the molecular scale, the exact process of feeling may conceivably occur at this scale (at the maximum). The Architecture, which is a framework or meta-solution, makes no demand one way or the other.

The Quale Garden

If a quale for Red really were fundamental in the universe, then there would have to be hundreds or more of qualia for all the other qualia- the taste of wine, the sound from a musical instrument and so on. Quite obviously this is literally unnatural. The grand "plan" in every case to date has been found to be a few simple fundamental elements, then placed in combination. This means the typical quale is actually a composite structure, that is, the "atoms" of the mind are not the "atoms" of the universe. So, we do need something smaller than a quale, or alternatively, some of the qualia are more elementary than others. At bottom will be the smallest experiences in the universe (the psybits).

The usually cited qualia include a cast that are deeply dissimilar, not peers.

Of special note are the differences between, for example, Red and Pain. First of all, there are a great many qualia (especially in principle) just as "elementary" as Red, which immediately suggests it is not a truly elementary feel. Second, there is nothing special about Red from the standpoint of the universe; again, it is just one arbitrary spot in a large field of characteristics (even though as it happens our nervous systems are specifically sensitive to it). Third and most saliently, Red has no intrinsic valence; Red is not intrinsically a painful thing or a bad thing or a good thing, it is just a factual condition.

Pain is entirely different in all three of these ways. There is pain and there is pleasure and there is nothing else in that family. Pain (or pleasure) is a very special occurrence in the universe, different from others. It has valence, and furthermore no factual content, only valence. The view that pain and Red are atomic siblings is false. Red and Blue and Shrill and Motivated are all of them more complicated than Pain and have internal structure. *To a mind*, or particular minds, they may be the simplest and most individual things detectable at the bottom, but that does not mean they are not made from simpler things (and they are).

In fact, the only clearly elemental qualia are pain and pleasure, the "emotional" ones. It turns out this is a profound observation and an important clue. Pain and pleasure are must-haves in the base feeling repertoire; no others are, they are constructed. The key categories of qualia, then, are Valent or Not, and Level of Complexity.

If a quale is a feel, then the simplest quale is the simplest feel. The simplest feel contains nothing other than feeling. This means a simplest quale cannot be about something, for then you have the feeling plus the thing about, and so here we encounter a difference in structural levels (levels of complexity).

To a mind, Red may be as elemental as pain, because inside a mind you are sitting on top of everything needed to begin making one. This does not mean Red is as simple as pain inside the fabric of this platform. Indeed, this has been one of the primary confusions historically in understanding consciousness.

Even at the lowest level, one can't get from an Is (information) to an Ought (which requires value and therefore feeling); the two different elements must become associated. The connection of feeling to information flows is ultimately what makes ideas "register." The idea is cognitive, the registration is feeling.

Necessary Qualities of Fundamental Consciousness

Qualia are the molecules of the mind, but not its quarks. And since qualia must be constructed of something similar (for workable emergence), this constructability becomes an essential feature of a successful architecture.

The answer is provided by the physical *field*, the mechanism that addresses another key problem, the fusion of feeling. This may be either an actual physical field, which we presume is the simplest explanation, or in principle another mechanism that is different but very similar in character.

We address first the build-up from fundamental pain to quale-level (mind-level) pain.

Note that by pain we mean "the painfulness of pain." Pure pain is the dysphoric ("emotional") aspect, the negative experience that is the end result of everything that causes pain.

If you are in slight pain and it gets much worse, we know that you will be experiencing the effect of more neurons activating, not just one or two becoming tremendously more active. At a physics level, the effect is many more psybits (fundamental feeling events) occurring, generally in close quarters. But what you *feel* at the mind level is not 1,000 individual small dots of pain. Your mind experiences the *same, unified thing, greatly amplified*. This means there must be a mechanism adding and joining these psybits into the whole pain. Note that simply interconnecting these feeling spots with neural information signals is not enough- that would still be separate little feelings, possibly now synchronized in time but still isolated as feelings and linked only by informational bridges. An additional mechanism is needed.

Physical interactions are all the behavior of forces. In a force field, every spot is subjected to the influence of every other spot in the same field. What results is a three-dimensional (four, including time) intensity cloud of a specific compositional shape. With this fusion, the otherwise freestanding feels join into a larger cloud of feeling. This is why you feel one big pain instead of a thousand little ones.

Pain is special as a feel/quale because pain is pain, on a small scale or a larger scale. Pain flows up from Layer 1 to Layer 2 in a rather direct and simple fashion.

Next we will discuss the informational qualia, which are very different. In elegant, Occamian economy, the same mechanism underlies these. The field mechanism is needed in the Consciousness Architecture.

Consciousness Architecture Layer 2

Unification of information can occur in one way by the interconnected firing of informational components, but that does not explain a feeling of Red that is both unitary and different from a feeling of Blue.

One will come to know a feeling of Red through very long experience with it, but as has been oft-cited, exactly what red is like to me may not be identical to what it is like for you. What is important in the fabric of mind is that it feels like *something* and it feels *different* from Blue, and from early on we then take these things (feelings) for granted as our minds continue to build.

Separate occurrences are not enough. *Both for feeling and for informational meaning* (which suggests the same mechanism is at play), we must have both the individual sub-aspects and unification of those into a whole. Assuming that there is only one level- qualia- is one of the fundamental errors in traditional thought on the topic, even before we get to the assumption that these individual qualia are the same thing as physical acts rather than compositions of physical acts.

We are provided an excellent clue toward the solution of the problem with pain (and pleasure). Pain is all the same, at all scales. It contains no information; it is purely valent.

Once we understand the necessary nature of fundamental consciousness, and that this is not the same as qualia, the next question presenting itself is how do we get from the former to the latter?

Meta-Symbols

To make thinking requires unique "symbols," things we can manipulate that represent other things. Even a symbol is divisible, and that is even before we deal with a symbol having both a name and an appearance. What is needed to start are unique identifiers- tags (meta-symbols)- to associate with large clouds of meaning. The tags, which are themselves each feeling clouds, become associated with yet more clouds, for its name, for its written visual representation if any, plus possible explicit association with net feelings of goodness or badness.

This assemblage provides two things: feeling and unique pattern of real things in space. One such blob "feels different" from another, both in the sense of the blob doing its own feeling and of interacting blobs "experiencing" it (which is the underpinnings of mind and which is two peers- being feeling- affected by each other).

The little clouds *feel different*. This, ultimately, is why Red is different from Blue and a hundred other "raw" feels. Red and blue must feel different, and we must have a means by which this can be so. This issue is right at the nexus between the cognitive and the emotional mental realms. It is not enough to be different informationally.

Different patterns of feeling in the clouds allow for the existence, recognizability, and reference of tags that are different. By themselves, they (the tags) have no meaning or clear symbolic identity, the purpose is simply to be different and distinguishable- unique identifiers. Precise visual or aural symbols can then exist- which themselves have some complexity- and then be attached to these tags, just as the many specific experiences that make up the meaning are attached to them.

What does "attached" mean? The attachment *to* both the symbolic detail and the meaning detail can be just cognitive- neural links that don't, themselves, have to feel. This is the connection between the what-it-is-like for the remembered meaning (and similarly for the visualization of the symbol) to the what-it-is-like of the tag.

These 3 identifiable clouds of experience will bump up against each other in the overall continuous train of feeling being created in the lower brain, while the higher has the parallel cognitive conversations, to whatever relative degrees the content has high cognitive content vs. high impact.

Each part of each cloud is being stimulated by a cognitive circuit (possibly by way of an additional connected neuron whose only function is to tweak the field), and then unified by the natural field(s).

Cognition evolves to where these differentiated and distinguishable tags become persistently associated with much larger "clouds" of cognitive activity. With these handles, entire ideas can be effectively referenced and manipulated.

This is what "meaning" is to us. Meaning as a term in common use means: 1) a collection of informational associations and 2) subjective experience of it. This is the blossoming of "what it is like."

Information is patterns of things, as with an abacus, and in full reduction the only things needed are the feelings. Anything else is intermediary, inert in the sense of importance to the abstract theory and architecture, other than to help arrange the patterns (which is where the wiring of the brain and physical shapes of neurons themselves come into play).

Random variability in the structure of neural nets and neurons yields unique, distinguishable signatures of the clouds of psybits mutually-influencing across space where these activations are happening in the brain. Red feels like something and Blue feels like something and they are different.

The full picture of this is feeling stimulated in the field by informational circuits and a return path (otherwise feelings would have no effect), and this occurs both very far down and at multiple levels, ascending with the size and sophistication of thoughts (bigger concepts will literally be larger than tiny concepts). There is a very intimate and deep back and forth of feeling what is known and happening and knowing what is being felt that forms the stuff of our minds (so much so that it has been difficult for us to tease these two strands apart).

It is necessary to have physics at work up to the point of constructing qualia. Qualia are the link from the actual to the abstract. Qualia are a conceptualization of the reality of feelings. With the "hardware" realization in place, we can manipulate and trade in the constructs conceptually, and the rest of the stack becomes "software" (and connective wiring). An important transition is made at this point: feeling is absolutely critical to how our minds works and this must be understood. Once it is, though, feeling can be considered a known aspect and discussed and used in an abstract way in recognizing a working architecture that relies upon it.

We note that feeling provides “information” of a sort, but with important difference from ordinary information. Feeling enables a key portion of the control section that guides the operation of the data paths and computation, and this is going on at all levels of the mind (though not in all portions).

The difference can be understood in terms of implications. Knowing of either kind can be *encoded* in information, which is then read so as to know what is in the message. The knowing elicited may be of a fact or of a pain/pleasure experience. The two are different, even though both may be represented and, in that way transported, by information.

Another way way to see the distinction is: Given enough book knowledge to read and enough mental ability and time you might understand any fact presented. However, to know what an experience was actually like can only be successfully transmitted to someone who has had the same experience or similar experiences. Otherwise one must resort to knowing factual features about the experience and cogitating about a described experience instead of being able to emulate personally the experience of the sender, to sympathize.

The top two layers of the architecture presented do not discuss physical reality, they discuss the structure of our minds. The bottom two layers explain what the fundamentals are for the top two layers to work with, and make the connection to the bottom of what we can introspect to, and call for specific explanation to ultimately prove those things at that bottom are the atoms of mind, validating the assumptions made by the higher two layers.

For the top two layers it is more dangerous to a full theory to be wrong about what is actually happening. But for the bottom two layers the most useful step is a characterization of the class of qualifying physical solutions, rather than the true single explanation. Correlation experiments may prove the class. Later knowledge will show the exact answer. This is the use of the very gift of abstraction that we have been discussing.

The stratification into layers correlates nicely with the different disciplines; neuroanatomical findings, for example, are especially useful throughout Layer 3, and harder to use in explanation beneath or above.

We have seen that cognition manipulates information and so is not of necessity related to the stuff that encodes it. Feeling is not about information and has everything to do with actual physical process. This means in its simplest form, mind needs nothing more for the stuff than the feeling, and the wiring that helps pattern the occurrences of the feeling; no other stuff is strictly required other than to maintain life and the mechanics and housekeeping of biological functioning.

This leads (distantly) to a subbranch of the present thinking that is (much) more speculative but not devoid of merit, involving the belief by some that life itself involves the fundamental consciousness force. The primary rational support for this notion, which probably originates from a "transcendent" suspicion (as did also that we are related to animals and stars), derives from the realization that feeling had to be discoverable by evolution. One possibility, then, is that it was discovered quite early by evolution and has been used to enable life processes that speak little (at least, obviously) to its experiential nature, just as electromagnetism and gravity have been used early and often. Note that base consciousness is expected by this paper to be variations in a field present everywhere. From the other direction, it is also possible that our particular region of space was particularly seeded not only with proto-life elements but with proto-consciousness elements.

Self

There is a self- the one- before awareness of self. When do things enter one's consciousness? When the chief, large, dominating (though shifting) cloud of feeling with embedded cognition that we call Me is affected by other clouds- when they come into effective contact because of out of a background of feeling noise there is a local strengthening that bridges into the primary Me. The Me is already a large cloud of assembled feelings, that has a (fuzzy) perimeter- a cloud that hangs together with only thin "mist" of disorganized consciousness between it and other little clouds.

To enter your conscious mind, you must feel what is entering. Not to feel is not to be conscious of.

To be conscious of is not the same as the core mechanism of attention, which is a cognitive mechanism. What you feel will in large part (the other part being cognitive things like difference) determine what the attention circuits bring into focus, and the cognitive things (and feelings, too, as objects of attention) then amplified rather than suppressed will in turn result in feelings, which will then affect attention, but the attention proper is a cognitive device. Feeling occurs in its usual independent and passive way, yet determines and thus drives the direction of thought.

Note that there is probably one self in each hemisphere of the brain. Your informationally linked but separated hemispheres may have distinct emotional lives, in keeping with some current theories. This also supports a prediction about the field strength formula of the consciousness force: it should feature high influence in the micron-millimeter scale and poor influence at the centimeter-meter scale.

Consciousness Architecture Layers 3 and 4

Once we have sentience- qualia- the stage is set to begin to actually build minds. Layers 3 and 4 are mind architecture. The internals of 3 and 4 are beyond the scope of this paper. There is much written of a preparatory nature on these speculations, which are being well-advanced by neuroscience.

The bootstrap gets us from Level 1 to Level 2. Evolutionary *cognitive* progress, continuing to exploit the emotional mechanism, gets us from there all the way up to chimpanzees, and internalized formal language, which is again a purely cognitive function, completes the stack.

Fortunately, it does not matter what bottom process is ultimately at play in formulating the architecture. Fundamental action X in the physical universe plugs in at the bottom and all the ascending logic holds without change (and does so even if the "fundamental" consciousness act is somehow slightly divisible or a small combination of things). Nonetheless, we later consider the specifics.

Cognition happens when rather than just link an informational and a pain/pleasure input to a (possibly reinforced) action to navigate in and/or manipulate the physical world, internal “objects” crafted of these sensations become themselves things to navigate to or around or, especially, to manipulate. In other words, cognition represents the birth of abstraction.

At the top of Layer 2, qualia that have been automatically generated appear from below and may be used for manipulation on the above side. There is also a what-it-is-like to *do* something on the above side, and the demands for action are sent down for their execution by unconscious inter-cooperative actions.

As to language and the internals it reveals, it interesting to note that Spanish, for example, features different words both for “to be” (fundamental essence, versus current state of affairs, which is higher) and “to know” (a fact, which is cognitive, versus a person, which is largely emotional).

Behind the forehead, what is known and what is wanted collide to result in action decisions (or at least awareness of them).

Evolvability

At this point we have an architecture, which stands for study irrespective of the specific explanation as to how it is "implemented" by Nature. But as evidence of its truth, it must be a possible destination of evolution.

The older and deeper parts of mammalian cortex form the connections to pain and pleasure that guide the cognitive circuits (literally) above. With this basic plan in place, the stage was set to scale up- with the same basic brain architecture- to more and more, and more refined, cognitive mass. Summary good/bad/pain/pleasure/warmer/colder is a very powerful concept key to advanced brains. As the ability to manipulate objects became more abstract, language first emerged and then became not just a communication feature but integral to human thought; communication itself became internalized and reflective; metalanguage (grammar, new notations, compound constructed words, and descriptive theories) emerged and itself became introspectively manipulatable.

Feeling continues to guide cognition at all levels. This allows three-pound brains to handle situations usually well enough that would be impossible to handle with combinatoric perfection, and provides an edge to conquer, in evolution, competitors and challenging niches. With bacteria and others optimized for the easier (in terms of complexity) niches, we populate the harder ones, requiring large-scale sophistication.

Getting Specific

Within the gross metatheory we have described are possible specific theories. Individual derivative theories may succeed in the lab or may fail, without necessarily disturbing the metatheory.

Why do we not detect the underlying things of which we have spoken? Following are among the more plausible possibilities.

Possibility 1: Fundamental feeling is another aspect of electromagnetism (or a combination of E-M and gravity) of which we are not aware. (Depending on the historical sequence of discovery, it would similarly have been possible to come to believe in the photon *before* realizing its connection to magnetism).

Possibility 2: Much of the fundamental interaction is in other dimensions. "Additional" dimensions arise often in physics theorizing.

Possibility 3: Consciousness is something we do already know, but we don't realize it yet. Dark Energy?

Possibility 4: An extension of the above is that *all* interactions are conscious and/or all energy is consciousness, but usually scattered and disorganized or differently organized.

Possibility 5: Less likely is that the consciousness particle(s) have not been found yet because they are too heavy. Some of the simplest theories require consciousness bosons with rather long range for fusion in the field, which advocates against heaviness. Still, relatively simple theories would have heavy particles. A slight elaboration might involve intermediary "messenger particles."

Note that there are many candidates to be consciousness at the fundamental level. A known or unknown particle's relative position, travel, acceleration, spin, spin change, precession and/or return, entanglement/disentanglement/wave function collapse, etc. etc. Any of these things that we have already observed could turn out to be the conscious act.

With meta-architecture in place we can look at what thinking may exist as to what evolution in our universe has implemented as the actual. (Note that the incorrectness of these architectural instances does not in any way invalidate the meta-architecture that is the subject of this paper.)

The most nominal model, regardless of effect- consciousness or other- is of a force charge, carrier particle having the charge, and mediator particle of the force. This leads to: "sentonic" charge, carried by potentially any particle (typically matter), and mediated by the "senton" (this is the Sentonic Theory of Fundamental Consciousness (STFC) (Sipfle, The Nature of Consciousness, 2018), (Sipfle, The Nature of Fundamental Consciousness, 2018)). It is noted that any or all of these three model elements could be things we already know. Perhaps gravity moves planets on the large scale and also explains tiny tag clouds of feeling at the small scale, for example.

For a fundamental consciousness source, Pepperell (Pepperell, 2018) speaks of "energy" very generally and has taken some first steps toward the direction we have discussed here. Notably he has taken an interdisciplinary and high-level approach. Schiffer (Schiffer, 2019) offers similar work from a medical perspective.

Poznanski (Poznanski, 2019) sees his equivalent to psybits arising at a slightly higher scale (molecular), and relies on quantum effects. This could be, although we have never seen such a basic phenomenon appear (correlate to physical action) higher up like that. All understood phenomena- including emergent phenomena of complex systems- are understandable in terms of fundamental particles and their interactions.

Poznanski offers one of the most complete (and thus most speculative) theories that is at least somewhat convincing. One of the most important individual parts to his work is the included metatheory that feeling emerges near the boundary of the quantum and the classical, with no new physics. In other words, to that author, a) feeling is an aspect of existing known processes, and b) its appearance is higher up than are other known phenomena- (bio)molecular-scale rather than particle-scale. For this reason, he sees the problem as one of biology rather than only of physics- a view that there is not enough in place for the process to happen until you have assembled biological material. To this end he espouses a philosophy called panexperiential materialism, in which consciousness requires certain "matter" and cannot exist without it. The chief deviation from this paper's view is that the matter is biological material rather than the physicist's matter, which means particles. Above the molecular scale the theories resemble each other: feeling in place as a physical process and deeply intertwined with the informational.

The mind is essentially a unified stage (or bottle) of thought and feeling, notwithstanding its ebbs, flows, fragmentation, and subconsciousness underpinnings. That stage is made of two things: the expanse and extent of neural hardware and the fields affected by its microscale movements. Topology does not create feeling (though it may create informational attention and self-referential "awareness," which is not the same as *feeling* aware or feeling an awareness of things). All the theories referenced in this section rely, expressively or tacitly, on field qualities.

Purely information-processing approaches to building artificial minds are unlikely to work for reproducing (and then amplifying) something like ourselves. Even if *represented* informationally, the concept of *value* summary interwoven at all levels of cognition is necessary to produce (simulate) something like our whole minds, which work by approximation and data reduction and parallel search and competition at many levels, guided by value.

A goal of this paper is to show the emotional base needed from which the then purely cognitive development of mind can proceed- the architecture of the fabric and platform atop which the architecture of the cognitive is then laid.

The problem breaks neatly into the platform (Layers 1 and 2) and the cognitive architecture atop it (Layers 3 and 4). The platform is itself neatly (and critically) divided into two steps- the fundamental action (the "transistors of the mind") and the means of their combination into the elements at the bottom of a cognitive mind (the interface to the operating system).

Once qualia become possible with that interface, the second architecture (Layers 3 and 4) will itself likely have a bottommost portion ("middleware layer") that manipulates the simplest things of a mind. Development of the mental middleware theory will likely accelerate the development of the rest above. For example, attention is a cognitive process.

The upper architecture then need only operate in a world of (meta)symbols, valences, and actions, and weave these together properly. This higher architecture need not concern itself with how anything below it actually works.

Looking ahead a little, the Middleware will start to categorize and label and compile into sequences and assemblies and place gross values on assemblages, and has the opportunity to create initial contacts to language even though that will come in full form much later in the ascension. This would bear a loose association both to animal cries and the first fuzzy prototypes of concepts in animal minds.

Forward Work

The layered taxonomy and architecture laid out in this paper provide an expected reality and a path for study. There are four neat boxes, each of which needs to be understood and each of which represents a field of specialty in solving consciousness. The snap-together layers may each be solved by different parties while still true to the architecture. To know completely what our minds really are, one has to know each of these layers, and the nature of the puzzle is different in each.

This applies to both theory and in actually building a mind. As with any good architecture, mind engineering becomes a layered process where each can proceed independently, with Layer 1 as the "silicon," Layer 2 as the fixed "firmware," and the rest as cables and software (which with pathway reconfigurability becomes all software in every way).

Once all of these layers are in place, a fifth layer can commence, which assumes thought capabilities and is data-driven, knowledge-driven, and hypothesis/abstraction driven, where large knowledge models are built and manipulated (there is also separately the opportunity for very rich emotional experiences, both by discovery and exploration and by intentional construction and acquisition, which is a different endeavor).

All of this understanding gives us the first clues to creating artifacts that *actually* feel. By fusing with them, this leads to the possibility- indeed, the likelihood- of extended life of individual human minds (not just of copies of them).

Summary

Electromagnetism is already heavily used in making up the fine informational computation and transmission patterning in the brain. We need something "out of band" from that for those to connect with to provide the emotional value.

Clouds of odd and differing shapes in a mist of consciousness "vapor" provide both valent experience as their essence and distinguishable larger-scale feelings that serve as meta-symbols (tags) around which clear and distinct mental objects can be constructed and then manipulated, like snowflakes around a mote of dust.

This paper has dealt with the necessity of explaining the *beginnings* of our minds in two steps:

4. The fundamental capacity in this universe to feel anything, at its most basic and indivisible.
5. The combining of this into the lowest level tier of experience that we can find introspectively from within our minds. This is the bootstrap from Layer 1 to Layer 2.

1 and 2 in place can then be assumed and provide the conceptual atoms and form the complete platform- the interface- upon which minds can be built, in layers/stages of cognitive capability.

With 1 and 2 in place there is sentience, and the mind from there becomes simply the result of increasingly fancy computation with connection to value, and self-observation, focus, and manipulation. Call that step 3. Step 3 is proceeding apace and is not the hardest problem today.

This platform of Layers 1 and 2 can be simulated with effectively exact fidelity, even without knowing the precise physical interaction used by our universe, allowing artificial "consciousness" which does not actually feel but is just as efficacious. (The model also allows real man-made consciousness to be inserted when that is mastered.)

Intelligence is the ability to solve problems. Intelligence is about factual situations and "cognitivizes" good/bad/pain/pleasure into, effectively, numbers. So, an equivalent form of the base that does not actually feel does not reduce the effectiveness of an artificial intelligence that has all the same knowledge as a person. The upper layers of the stack are fabric-independent.

We must have stuff to move around in order to represent and manipulate information. Computationally, it doesn't matter what the stuff is. Any stuff that exists in self-local little pieces can be arranged.

We need a certain specific thing to be feeling- specific stuff. Feeling by itself on the local level has nothing to do with information, nothing to do with arrangement.

The fundamental magic of our minds is the marriage of information processing with feeling, and in particular, valent feeling. The information is the arrangement of stuff. The stuff is feeling.

We see:

- We need stuff that feels. Not just any stuff.

- We need stuff to arrange and rearrange. Any readily arrangeable stuff will do.
- The stuff can be entirely the same that is arranged, or there can be a somewhat more complicated arrangement between the two. Either way, as a practical matter, just a little way up it is the same stuff.
- Feeling goes way down. Valence goes way down.
- Feeling does two things- fuse/join and thus generalize and simplify into grosser states, and provide on the same many scales a sense and score of good/bad.
- Subjective feeling is an epiphenomenon is in the sense that any natural mechanism that had the utility just described would do. In our universe pain and pleasure are available for the task and have been harnessed.
- Electromagnetic processes are both coupled with and have a high degree of separateness from feeling. The separateness is to provide new utility; feeling does a "processing" that the electromagnetic processes are enhanced by, as proven by evolution weaving feeling in.

Only three basic things are needed to explain the mind

- Physics-level understanding of feeling.
- Explanation of path from physics level to bottom-most mind (qualia in a unified place)
- Cognitive processing mechanisms, ultimately including language. These are coming in to view

via neuroscience from microanatomy and various scanning and probing.

These things are reflected in the corresponding architecture of the mind:

5. Language and formal reasoning.
6. Cognitive processing, pre-language.

----

3. Building from physics to bottom of mind: integration of cognitive with feeling, formation of bottom brain experiences and actions.
2. Physics-Manipulable objects to represent information, feeling action to summarize and evaluate.

With 1 and 2 as the hardware and firmware, 3 and 4 are effectively all networking and software (even if much of the algorithmic is done through cellular processes in the case of a biological brain).

Where there has been much confusion is in 1 and 2, and this is what tells us what consciousness *is*- in other words the difference between a simulated consciousness and an actual consciousness. We have not yet detected the base constituents. As with many other advances in earlier times with thin data, we are driven to our conclusions by reason, and this time we know how Nature tends to work.

Meanings not only feel like something in an active mind, they are made entirely out of feelings, in that even the informational content is patterns- patterns of feelings. This brings together the informational and the subjective aspect of knowing.

While information/cognitive is the patterns and independent of the stuff, pain and pleasure are stuff of a particular kind and independent of information patterns. The two become associated early and often and form the fundamental fabric of what is our consciousness. As this dance proceeds higher, we know what we feel and that we feel, and we feel that we know and what it is like to know.

References

Pepperell, R. (2018). Consciousness as a Physical Process Caused by the Organization of Energy in the Brain. *Frontiers in Psychology*.

Poznanski, R. (2019). Theorising How the Brain Encodes Consciousness Based on Negentropic Entanglement. *Journal of Integrative Neuroscience*.

Schiffer, F. (2019). The Physical Nature of Subjective Experience and Its Interaction with the Brain. *Medical Hypotheses*.

Sipfle, K. (2018). *The Nature of Consciousness.* New York: Barnes & Noble Press.

Sipfle, K. (2018). The Nature of Fundamental Consciousness. *ResearchGate*.

Sipfle, K. (2019). The Solvability of The Hard Problem of Consciousness and its Relation to Fundamental

Consciousness and Human Consciousness. *ResearchGate*.

Support for the Sentonic Theory of Fundamental Consciousness (STFC) in Recent Findings Supporting Ephaptic Coupling

Karl Sipfle, Independently

at NASA GSFC

ksipfle@umich.edu

Abstract: The Sentonic Theory of Fundamental Consciousness is a theory of an underlying fundamental consciousness operating at the physics level. Research has demonstrated certain expectations of that theory, namely that ephaptic neural communication does occur, and that the cause is the electric field variations generated by operating neurons. The electric field variations may be part of a multistep interaction chain. Weak electrical activity was specifically found to influence pyramidal cells, the most popular candidates for carrying consciousness. The cited research and the Sentonic Theory of Fundamental Consciousness have also both independently led to speculations of an important role for the dense neuron packing and/or regular formations of axons in some brain tissue.

Keywords: Sentonic Theory of Fundamental Consciousness, STFC, consciousness, brain, mind, ephaptic, electromagnetic, field

[1] describes a theory of an underlying fundamental consciousness operating at the physics level, which we refer to here for continuing reference as the Sentonic Theory of Fundamental Consciousness (STFC). The described theory makes, among others, these (independent) predictions.

1. There is an extra-synaptic (ephaptic) means of communications between neurons.

2. (In order to most simply explain raw, subjective feeling as well as its fused, singular nature) this communication necessarily passes bidirectionally through a force field, a new ("sentonic") field.

3. The most likely link from the familiar operations of the brain to this new field is the electromagnetic (E-M) action of operating neurons.

[2] demonstrates, among other things:

1. Ephaptic neural communication does necessarily occur.

2. "The" cause and medium are the electric field variations generated by operating neurons.

But the view of other researchers [3] that the weak electric fields are thought to be too small to produce excitation directly may also be true. In STFC, E-M activity is our starting point of transmission, which is interlinked with sentonic activity (and sentonic fields contain energy).

Without further inspection, observed E-M activity may look to be the sole agent, but that is not strictly proven.

Further, *something* explains fundamental consciousness. All paths of communication are suspects for study, certainly including the discovered ephaptic path.

The slow waves that are the subject of Chiang [2] are found throughout the neocortex- possibly originating in the prefrontal- as well as the hippocampus. Weak electrical activity has specifically been seen influencing the pyramidal cells [4], which are likely the conscious ones [5].

While Chiangs's evidence does not constitute proof that the entire Sentonic Theory is correct, it does support the major (and controversial) prediction of ephaptic coupling, and causally associates that with neural electrical activity, as both predicted by STFC.

In the course of [2], and starting from an entirely different point, two observations are made that are also found in [5]. The first is the observation that some brain tissue stands out from the rest in being of densely packed cells and/or axons in regular formations; [5] speculates the distinction may play a role in effectiveness in interacting with the sentonic field (possibly as a second-order effect by organizing causative E-M field variations). The second observation is that the concepts noted here may explain what is actually happening in transcranial stimulation therapies.

Conclusion

Chiang's work [2] supports key and controversial predictions of the STFC and produces no contradictions to it.

References

[1] K. Sipfle, "The Nature of Fundamental Consciousness (prepirnt)," *PsyArXiv,* 2018.

[2] C.-C. Chiang, "Slow periodic activity in the longitudinal hippocampal slice can selfpropagate non-synaptically by a mechanism consistent with ephaptic coupling," *The Journal of Physiology,* 2019.

[3] P. R. Anastassiou CA, "Ephaptic coupling of cortical neuron," *Nat Neurosci,* Vols. 217-223, 2011.

[4] G. B. Francis JT, "Sensitivity of neurons to weak electric fields," *Neurosci,,* pp. 7255-7261, 2003.

[5] K. Sipfle, The Nature of Consciousness, New York: Barnes and Noble Press, 2018.

www.ingramcontent.com/pod-product-compliance
Lightning Source LLC
LaVergne TN
LVHW020325220926
848484LV00001BB/109